Stem CELLS
and the FUTURE OF REGENERATIVE MEDICINE

Committee on the Biological and Biomedical Applications of
Stem Cell Research

Board on Life Sciences
National Research Council

Board on Neuroscience and Behavioral Health
Institute of Medicine

NATIONAL ACADEMY PRESS
Washington, D.C.

NATIONAL ACADEMY PRESS   2101 Constitution Avenue, N.W.   Washington, DC 20418

NOTICE: The project that is the subject of this report was approved by the Governing Board of the National Research Council, whose members are drawn from the councils of the National Academy of Sciences, the National Academy of Engineering, and the Institute of Medicine. The members of the committee responsible for the report were chosen for their special competencies and with regard for appropriate balance.

This study was supported by the National Research Council Fund and by the Ellison Medical Foundation under Agreement no. NI-CW-0007-01. Any opinions, findings, conclusions, or recommendations expressed in this publication are those of the authors and do not necessarily reflect the views of the organizations or agencies that provided support for the project.

Library of Congress Cataloging-in-Publication Data

Stem cells and the future of regenerative medicine / Committee on the Biological and Biomedical Applications of Stem Cell Research, Commission on Life Sciences National Research Council.
    p. cm.
Includes bibliographical references and index.
 ISBN 0-309-07630-7
 1. Stem cells—Research—Government policy—United States. I. National Research Council (U.S.). Committee on the Biological and Biomedical Applications of Stem Cell Research.
 QH587 .S726 2001
 571.8′35—dc21

                          2001006360

Cover: Background image courtesy of Musee National du Chateau de Malmaison, Rueil-Malmaison/Lauros-Giraudon, Paris/SuperStock; stem cell photo courtesy of James Thomson Laboratory, University of Wisconsin Board of Regents.

Additional copies of this report are available from National Academy Press, 2101 Constitution Avenue, NW, Lockbox 285, Washington, DC 20055; (800) 624-6242 or (202) 334-3313 (in the Washington metropolitan area); Internet, http://www.nap.edu

Printed in the United States of America.

THE NATIONAL ACADEMIES

National Academy of Sciences
National Academy of Engineering
Institute of Medicine
National Research Council

The **National Academy of Sciences** is a private, nonprofit, self-perpetuating society of distinguished scholars engaged in scientific and engineering research, dedicated to the furtherance of science and technology and to their use for the general welfare. Upon the authority of the charter granted to it by the Congress in 1863, the Academy has a mandate that requires it to advise the federal government on scientific and technical matters. Dr. Bruce M. Alberts is president of the National Academy of Sciences.

The **National Academy of Engineering** was established in 1964, under the charter of the National Academy of Sciences, as a parallel organization of outstanding engineers. It is autonomous in its administration and in the selection of its members, sharing with the National Academy of Sciences the responsibility for advising the federal government. The National Academy of Engineering also sponsors engineering programs aimed at meeting national needs, encourages education and research, and recognizes the superior achievements of engineers. Dr. Wm. A. Wulf is president of the National Academy of Engineering.

The **Institute of Medicine** was established in 1970 by the National Academy of Sciences to secure the services of eminent members of appropriate professions in the examination of policy matters pertaining to the health of the public. The Institute acts under the responsibility given to the National Academy of Sciences by its congressional charter to be an adviser to the federal government and, upon its own initiative, to identify issues of medical care, research, and education. Dr. Kenneth I. Shine is president of the Institute of Medicine.

The **National Research Council** was organized by the National Academy of Sciences in 1916 to associate the broad community of science and technology with the Academy's purposes of furthering knowledge and advising the federal government. Functioning in accordance with general policies determined by the Academy, the Council has become the principal operating agency of both the National Academy of Sciences and the National Academy of Engineering in providing services to the government, the public, and the scientific and engineering communities. The Council is administered jointly by both Academies and the Institute of Medicine. Dr. Bruce M. Alberts and Dr. Wm. A. Wulf are chairman and vice chairman, respectively, of the National Research Council.

COMMITTEE ON THE BIOLOGICAL AND BIOMEDICAL
APPLICATIONS OF STEM CELL RESEARCH

BERT VOGELSTEIN (*Chair*), Johns Hopkins Oncology Center,
Baltimore, and Howard Hughes Medical Institute
BARRY R. BLOOM, Harvard School of Public Health, Cambridge,
Massachusetts
COREY S. GOODMAN, University of California, Berkeley, and
Howard Hughes Medical Institute
PATRICIA A. KING, Georgetown University Law Center,
Washington, D.C.
GUY M. MCKHANN, Johns Hopkins University School of Medicine,
Baltimore
MYRON L. WEISFELDT, Columbia University College of Physicians
and Surgeons, New York
KATHLEEN R. MERIKANGAS (liaison, Board on Neuroscience and
Behavioral Health), Yale University, New Haven, Connecticut

Project Staff

FRANCES E. SHARPLES, Director, Board on Life Sciences
TERRY C. PELLMAR, Director, Board on Neuroscience and
Behavioral Health
ROBIN A. SCHOEN, Program Officer
JANET E. JOY, Senior Program Officer
BRIDGET K. B. AVILA, Senior Project Assistant
LAURA T. HOLLIDAY, Research Assistant
DEREK M. SWEATT, Research Assistant
NORMAN GROSSBLATT, Senior Editor

Acknowledgments

T his report is the product of many individuals. First, we would like to thank all the speakers who attended our workshop, Stem Cells and the Future of Regenerative Medicine, on June 22, 2001. Without the input of each of these speakers, this report would not have been possible.

Iqbal Ahmad, University of Nebraska Medical Center
George Annas, Boston University Schools of Medicine and
    Public Health
Ernest Beutler, Scripps Research Institute
Kevin FitzGerald, Georgetown University
Fred Gage, Salk Institute
Margaret Goodell, Taylor College of Medicine
Marcus Grompe, Oregon Health Sciences University
Ihor Lemischka, Princeton University
Olle Lindvall, Lund University
Ron McKay, National Institute of Neurological Disorders
    and Stroke
Thomas Okarma, Geron Corporation
David Prentice, Indiana State University
Arti Rai, University of Pennsylvania School of Law
Jay Siegel, Food and Drug Administration
James Thomson, University of Wisconsin
LeRoy Walters, Georgetown University
Irving Weissman, Stanford University

Second, this report has been reviewed in draft form by individuals chosen for their diverse perspectives and technical expertise, in accordance with procedures approved by the NRC's Report Review Committee. The purpose of this independent review is to provide candid and critical comments that will assist the institution in making its published report as sound as possible and to ensure that the report meets institutional standards for objectivity, evidence, and responsiveness to the study charge. The review comments and draft manuscript remain confidential to protect the integrity of the deliberative process. We wish to thank the following individuals for their review of this report:

Fred Alt, Howard Hughes Medical Institute, Harvard Medical School
Fred Appelbaum, Fred Hutchinson Cancer Research Center
Daniel Callahan, The Hastings Center
R. Alta Charo, University of Wisconsin Law School
Carolyn Compton, McGill University
William Danforth, Washington University
Neal First, University of Wisconsin
Barbara Gastel, Texas A&M University
John Gerhart, University of California, Berkeley
Paul Gilman, Celera Genomics
Micheline Mathews-Roth, Harvard Medical School
Martin Raff, University College London
Nathan Rosenberg, Stanford University
Evan Snyder, Boston Children's Hospital
Virginia Weldon, Monsanto Company

Although the reviewers listed above have provided many constructive comments and suggestions, they were not asked to endorse the conclusions or recommendations nor did they see the final draft of the report before its release. The review of this report was overseen by Ronald Estabrook of the University of Texas Southwestern Medical Center and Floyd Bloom of the Scripps Research Institute. Appointed by the

National Research Council, they were responsible for making certain that an independent examination of this report was carried out in accordance with institutional procedures and that all review comments were carefully considered. Responsibility for the final content of this report rests entirely with the authoring committee and the institution.

Preface

Stem cell research has the potential to affect the lives of millions of people in the United States and around the world. This research is now regularly front-page news because of the controversy surrounding the derivation of stem cells from human embryos. Realizing the promise of stem cells for yielding new medical therapies will require us to grapple with more than just scientific uncertainties. The stem cell debate has led scientists and nonscientists alike to contemplate profound issues, such as who we are and what makes us human beings.

The excitement and controversy surrounding stem cells caused the National Research Council's Board on Life Sciences and the Institute of Medicine's Board on Neuroscience and Behavioral Health to recommend that the National Academies sponsor a workshop to assess the scientific and therapeutic value of stem cells. The presidents of the National Academies agreed and provided most of the funding that supported the production of this report. The Ellison Foundation provided additional funding.

In a collaboration of the two boards, the Committee on the Biological and Biomedical Applications of Stem Cell Research was formed. The persons appointed to serve on the committee have a wealth of expertise in the basic and clinical biomedical sciences but do not themselves perform stem cell research. The latter characteristic was intended to ensure that none of the committee members had a vested interest in any form of stem cell research. Expertise represented on the committee includes molecular biology, immunology, cell biology, cardiology, hematology, neurosciences, developmental biology, infectious disease, cancer, and bioethics, all of which are integrally related to stem cell research and its potential for developing tissue-replacement therapies that will restore lost function in damaged organs.

At the committee's workshop, held on June 22, 2001, scientists, philosophers, ethicists, and legal experts presented their views in two general categories. First, leading scientific investigators addressed the following scientific questions: What are stem cells? What are their sources, and what biological differences exist among cells of different origins? How do these differences translate into advantages or disadvantages for research and medical applications? What is the potential of stem cells for regenerative medicine, and what obstacles must be overcome to make them useful for new medical therapies? Second, experts in philosophy, law, and ethics presented a variety of ethical and other arguments relevant to public-policy considerations on stem cells. Audio files of the speakers' presentations are available until December 31, 2002, at the workshop Web site: www.nationalacademies.org/stemcells.

This report presents the committee's findings and recommendations. It is based on careful consideration of information presented at the workshop and on data and opinions found in the scientific and other scholarly literature. The committee is extremely respectful of all perspectives in this debate and has taken them into account in forming its recommendations.

I wish to thank all the members of the committee for their valuable contributions and especially for their insights into both the scientific and the societal issues. In particular, Corey Goodman, chair of the Board on Life Sciences, was responsible for much of the initial impetus for the workshop. I also wish to acknowledge the staff of the National Research Council (Robin Schoen, Bridget Avila, and Fran Sharples) and the

Institute of Medicine (Janet Joy and Terry Pellmar) for their thorough, thoughtful, and efficient assistance with all aspects of the workshop and report preparation. This report would have been impossible without them.

> Bert Vogelstein, Chair
> Committee on the Biological and Biomedical
> Applications of Stem Cell Research

Contents

Stem CELLS

and the
FUTURE OF
REGENERATIVE
MEDICINE

Executive Summary

S tem cell research offers unprecedented opportunities for developing new medical therapies for debilitating diseases and a new way to explore fundamental questions of biology. Stem cells are unspecialized cells that can self-renew indefinitely and also differentiate into more mature cells with specialized functions. Research on human embryonic stem cells, however, is controversial, given the diverse views held in our society about the moral and legal status of the early embryo. The controversy has encouraged provocative and conflicting claims both inside and outside the scientific community about the biology and biomedical potential of both adult and embryonic stem cells.

The National Research Council and Institute of Medicine formed the Committee on the Biological and Biomedical Applications of Stem Cell Research to address the potential of stem cell research. The committee organized a workshop that was held on June 22, 2001. At the workshop, the committee heard from many leading scientists who are engaged in stem cell research and from philosophers, ethicists, and legal scholars. (Audio files of the speakers' presentations are available until December 31, 2002, at the workshop Web site, www.nationalacademies.org/stemcells.)

The participants discussed the science of stem cells and a variety of ethical and other arguments relevant to public policy as it applies to stem cells. The committee considered the information presented, explored the literature on its own, and contemplated the substance and importance of the preliminary data from recent stem cell experiments. The committee's deliberations on the issues led to the following conclusions and recommendations.

- Experiments in mice and other animals are necessary, but not sufficient, for realizing the potential of stem cells to develop tissue-replacement therapies that will restore lost function in damaged organs. Because of the substantial biological differences between nonhuman animal and human development and between animal and human stem cells, studies with *human* stem cells are essential to make progress in the development of treatments for *human* disease, and this research should continue.

- There are important biological differences between adult and embryonic stem cells and among adult stem cells found in different types of tissue. The implications of these biological differences for therapeutic uses are not yet clear, and additional data are needed on all stem cell types. Adult stem cells from bone marrow have so far provided most of the examples of successful therapies for replacement of diseased or destroyed cells. Despite the enthusiasm generated by recent reports, the potential of adult stem cells to differentiate fully into other cell types (such as brain, nerve, pancreas cells) is still poorly understood and remains to be clarified. In contrast, studies of human embryonic stem cells have shown that they can develop into multiple tissue types and exhibit long-term self-renewal in culture, features that have not yet been demonstrated with many human adult stem cells. The application of stem cell research to therapies for human disease will require much more knowledge about the biological properties of all types of stem cells. Although stem cell research is on the cutting edge of biological science today, it is still in its infancy. Studies of both embryonic and adult human stem cells will be required to most efficiently advance the scientific and therapeutic potential of regenerative medicine. Moreover, research on embryonic stem cells will be important to inform research on adult stem cells, and vice versa. Research on both adult and embryonic human stem cells should be pursued.

•   Over time, all cell lines in tissue culture change, typically accumulating harmful genetic mutations. There is no reason to expect stem cell lines to behave differently. In addition, most existing stem cell lines have been cultured in the presence of non-human cells or serum that could lead to potential human health risks. Consequently, while there is much that can be learned using existing stem cell lines if they are made widely available for research, such concerns necessitate continued monitoring of these cells as well as the development of new stem cell lines in the future.

•   High-quality, publicly funded research is the wellspring of medical breakthroughs. Although private, for-profit research plays a critical role in translating the fruits of basic research into medical advances that are broadly available to the public, stem cell research is far from the point of providing therapeutic products. Without public funding of basic research on stem cells, progress toward medical therapies is likely to be hindered. In addition, public funding offers greater opportunities for regulatory oversight and public scrutiny of stem cell research. Stem cell research that is publicly funded and conducted under established standards of open scientific exchange, peer review, and public oversight offers the most efficient and responsible means of fulfilling the promise of stem cells to meet the need for regenerative medical therapies.

•   Conflicting ethical perspectives surround the use of embryonic stem cells in medical research, particularly where the moral and legal status of human embryos is concerned. The use of embryonic stem cells is not the first biomedical research activity to raise ethical and social issues among the public. Restrictions and guidelines for the conduct of controversial research have been developed to address such concerns in other instances. For example, when recombinant-DNA techniques raised questions and were subject to intense debate and public scrutiny, a national advisory body, the Recombinant DNA Advisory Committee, was established at the National Institutes of Health (NIH) to ensure that the research met the highest scientific and ethical standards. If the federal government chooses to fund research on human embryonic stem cells, a similar national advisory group composed of exceptional researchers, ethicists, and other stakeholders should be established at NIH to oversee it. Such a group should ensure that proposals to work on human embryonic stem cells are scientifically justified and should scrutinize such proposals for compliance with federally mandated ethical guidelines.

•   Regenerative medicine is likely to involve the implantation of new tissue in patients with damaged or diseased organs. A substantial obstacle to the success of transplantation of any cells, including stem cells and their derivatives, is the immune-mediated rejection of foreign tissue by the recipient's body. In current stem cell transplantation procedures with bone marrow and blood, success can hinge on obtaining a close match between donor and recipient tissues and on the use of immuno-suppressive drugs, which often have severe and life-threatening side effects. To ensure that stem cell-based therapies can be broadly applicable for many conditions and individuals, new means to overcome the problem of tissue rejection must be found. Although ethically controversial, somatic cell nuclear transfer, a technique that produces a lineage of stem cells that are genetically identical to the donor, promises such an advantage. Other options for this purpose include genetic manipulation of the stem cells and the development of a very large bank of embryonic stem cell lines. In conjunction with research on stem cell biology and the development of stem cell therapies, research on approaches that prevent immune rejection of stem cells and stem cell-derived tissues should be actively pursued.

The committee is aware of and respectful of the wide array of social, political, legal, ethical, and economic issues that must be considered in policy-making in a democracy. And it is impressed by the commitment of all parties in this debate to life and health, regardless of the different conclusions they draw. The committee hopes that this report, by clarifying what is known about the scientific potential of stem cells and how that potential can best be realized, will be a useful contribution to the debate and to the enhancement of treatments for disabling human diseases and injuries. On August 9, 2001, when President Bush announced a new federal policy permitting limited use of human embryonic stem cells for research, this report was already in review. Because this report presents the committee's interpretation of the state of the science of stem cells independent of any specific policy, only minor modifications to refer to the new policy have been made in the report.

RECOMMENDATIONS

1. Studies with *human* stem cells are essential to make progress in the development of treatments for *human* disease, and this research should continue.

2. Although stem cell research is on the cutting edge of biological science today, it is still in its infancy. Studies of both embryonic and adult human stem cells will be required to most efficiently advance the scientific and therapeutic potential of regenerative medicine. Research on both adult and embryonic human stem cells should be pursued.

3. While there is much that can be learned using existing stem cell lines if they are made widely available for research, concerns about changing genetic and biological properties of these stem cell lines necessitate continued monitoring as well as the development of new stem cell lines in the future.

4. Human stem cell research that is publicly funded and conducted under established standards of open scientific exchange, peer review, and public oversight offers the most efficient and responsible means to fulfill the promise of stem cells to meet the need for regenerative medical therapies.

5. If the federal government chooses to fund human stem cell research, proposals to work on human embryonic stem cells should be required to justify the decision on scientific grounds and should be strictly scrutinized for compliance with existing and future federally mandated ethical guidelines.

6. A national advisory group composed of exceptional researchers, ethicists, and other stakeholders should be established at the National Institutes of Health (NIH) to oversee research on human embryonic stem cells. The group should include leading experts in the most current scientific knowledge relevant to stem cell research who can evaluate the technical merit of any proposed research on human embryonic stem cells. Other roles for the group could include evaluation of potential risks to research subjects and ensuring compliance with all legal requirements and ethical standards.

7. In conjunction with research on stem cell biology and the development of potential stem cell therapies, research on approaches that prevent immune rejection of stem cells and stem cell-derived tissues should be actively pursued. These scientific efforts include the use of a number of techniques to manipulate the genetic makeup of stem cells, including somatic cell nuclear transfer.

CHAPTER ONE

Project Overview and Definitions

This report addresses key questions about the biology and therapeutic potential of human stem cells, undifferentiated cells that can give rise to specialized tissues and organs. Medical and scientific interest in stem cells is based on a desire to find a source of new, healthy tissue to treat diseased or injured human organs. It is known that some organs, such as the skin and the liver, are adept at regenerating themselves when damaged, but it is not yet understood why and how some tissues have this capability and others do not. Recent research has indicated that stem cells are a key to these regenerative properties.

There are confirmed sources of stem cells in adult tissues, such as bone marrow, that maintain the ability to differentiate into the diverse cell types of that tissue throughout the life of an organism. However, cells that maintain the ability to divide and differentiate into more specialized cells of different tissue types are rare in the adult. In contrast, the seemingly unlimited potential of the undifferentiated cells of the early embryo has made embryonic stem cells the focus of great scientific interest. Since 1998, when James Thomson of the University of Wisconsin-Madison developed the first human embryonic stem cell (ESC) cultures, increasing attention has been paid to scientific reports hinting at the therapeutic potential of stem cells for treating various degenerative diseases and injuries (Thomson et al., 1998). What is now known as regenerative medicine seeks to understand how and why stem

TABLE 1. Potential US Patient Populations for Stem Cell-Based Therapies

The conditions listed below occur in many forms and thus not every person with these diseases could potentially benefit from stem cell-based therapies. Nonetheless, the widespread incidence of these conditions suggests that stem cell research could help millions of Americans

Condition	Number of patients
Cardiovascular disease	58 million
Autoimmune diseases	30 million
Diabetes	16 million
Osteoporosis	10 million
Cancers	8.2 million
Alzheimer's disease	5.5 million
Parkinson's disease	5.5 million
Burns (severe)	0.3 million
Spinal-cord injuries	0.25 million
Birth defects	0.15 million/year

Source: Derived from Perry (2000).

cells, whether derived from human embryos or adult tissues, are able to develop into specialized tissues, and seeks to harness this potential for tissue-replacement therapies that will restore lost function in damaged organs.

The list of diseases and injuries cited as potential targets of stem cell therapy reveals, in large measure, why stem cells offer so much hope for revolutionary advances in medicine (Table 1). Many of them—such as Parkinson's disease, diabetes, heart disease, Alzheimer's disease, and spinal cord injury—have few or no treatment options, so millions of Americans are currently looking for cures.

The hope of using stem cells to produce regenerative therapies poses fundamental questions: Do human ESCs hold all the clinical promise attributed to them? Is realization of that promise imminent? Do stem cells from all sources have the same abilities? What is their potential for regenerative medicine?

THE CHARGE TO THE COMMITTEE

Members of the National Research Council's Board on Life Sciences and members of the Institute of Medicine's Board on Neuroscience and Behavioral Health independently decided in December 2000 that they should sponsor a workshop on the scientific and medical value of stem cell research. The Committee on the Biological and Biomedical Applications of Stem Cell Research was appointed to organize the workshop and to produce a report on the biology and biomedical applications of stem cells in regenerative medicine. (Appendix A provides biographical sketches of the committee members.)

The charge to the committee was as follows:

> An appointed committee will organize a workshop on the biology and biomedical applications of stem cells. The workshop will examine several aspects of stem cell research, including: the biological properties of stem cells in general, the current state of knowledge about the molecular and cellular controls that govern transdifferentiation in cells originating from different types of tissues, the use of stem cells to generate neurons, heart, kidney, blood, liver and other tissues, and the prospective clinical uses of these tissues. The workshop will consider the biological differences of cells obtained from different sources, for example, embryos, fetal tissues, or adult tissues, and discuss concerns about the use of various sources of stem cells. The committee will produce a report that summarizes the workshop and the scientific and public policy concerns that present both opportunities and barriers to progress in this field.

The committee's workshop took place on June 22, 2001, at the National Academy of Sciences in Washington, D.C.; Appendix B contains the meeting agenda and biographies of the presenters. Audio files of the speakers' presentations will be available at the workshop Web site: www.nationalacademies.org/stemcells until December 31, 2002.

It is important to explain the limits of the committee's charge and work. Although data and opinions in the scientific and other scholarly literature were examined, the project did not attempt an exhaustive review of the scientific literature in this field. It should be noted that shortly after the workshop, the National Institutes of Health released a major report on the "Scientific Progress and Future Research Directions" of stem cells, and this document has provided valuable information for the committee's report (NIH, 2001).

The committee organized the workshop to address key issues in the status of stem cell research by gathering information from scientific leaders in the field. In addition, the workshop provided an opportunity for the committee to hear from both those who support embryonic stem cell research and those who oppose it on ethical grounds. The committee did not attempt to resolve the ethical dilemmas and limits its comments to scientific points intended to clarify or inform the ethical discussion. This report synthesizes the workshop presentations and puts forward the committee's conclusions drawn from that meeting. In particular, the report addresses the following questions:

- What characteristics of stem cells make them desirable for regenerative medicine?
- Which biological features of stem cells are well established? Which are uncertain?
- What implications do the biological features of different stem cells have for the development of therapeutic applications?
- What opportunities and barriers does stem cell research face, and how are they relevant to medical therapies?

The committee placed off limits the issue of reproductive cloning, which is sometimes linked to stem cell research because in both cases the somatic cell nuclear transfer (SCNT) technique can be used to create embryos (see Box). The interest in this technique for stem cell research is related to the possibility of producing stem cells for regenerative therapy that are genetically matched to the person needing a tissue transplant. The immune system is poised to reject tissue transplants

Comparison of Stem Cell Production with Reproductive Cloning

The goal of stem cell research using the somatic cell nuclear transfer (SCNT) technique must be sharply contrasted with the goal of reproductive cloning, which, using a similar technique, aims to develop an embryo that is genetically identical with the donor of its genes and then implant that embryo in a woman's uterus and allow it to mature to birth. Cloning for reproductive purposes will be the subject of a separate report now being developed by the National Academies' Committee on the Scientific and Medical Aspects of Human Cloning. In the table below, the cellular materials and techniques of stem cell research are compared to that of reproductive cloning.

	Adult and Fetal Stem Cells	Embryonic Stem Cells	Embryonic Stem Cells Produced with the SCNT Technique	Reproductive Cloning: Embryos Produced with the SCNT Technique
Purpose of use	To obtain undifferentiated stem cells for research and therapy	To obtain undifferentiated stem cells for research and therapy	To obtain undifferentiated stem cells that are genetically matched to recipient for research and therapy	To produce embryo for implantation, leading to birth of a child
Starting material	Isolated stem cells from adult or fetal tissue	Cells from an embryo at blastocyst stage produced by fertilization	Cells from a blastocyst produced by development of an enucleated egg supplied with nucleus from patient's somatic cell (SCNT technique)	Enucleated egg supplied with nucleus from donor's somatic cell (SCNT technique)
End product	Cells produced in culture to replenish diseased or injured tissue	Cells produced in culture to replenish diseased or injured tissue	Cells produced in culture to replenish diseased or injured tissue	Embryo derived from development of egg, implanted and allowed to develop to birth

from genetically non-identical people, and immunological rejection poses serious clinical risks that can be life-threatening. Overcoming the threat of immunological rejection is thus one of the major scientific challenges to stem cell transplantation and, indeed, for transplantations of any sort. The SCNT technique offers the possibility of deriving stem cells for transplantation from the recipient's own cells. Such cells would produce only the patient's own proteins and would not cause an immunological reaction when transplanted into that patient.

The committee is respectfully mindful of the wide array of social, political, legal, ethical, and economic issues that must be considered in policy-making in a democracy. And it is impressed by the commitment of all parties in this debate to life and health, regardless of the different conclusions they draw. The committee hopes that, by addressing questions about the scientific potential of stem cell and how that potential can be best realized, it can contribute usefully to the debate and to the enhancement of treatments for disabling human diseases and injuries.

WHAT ARE STEM CELLS? BASIC DEFINITIONS

Stem cells are unspecialized cells that can self-renew indefinitely and that can also differentiate into more mature cells with specialized functions. In humans, stem cells have been identified in the inner cell mass of the early embryo; in some tissues of the fetus, the umbilical cord and placenta; and in several adult organs. In some adult organs, stem cells can give rise to more than one specialized cell type within that organ (for example, neural stem cells give rise to three cell types found in the brain-neurons, glial cells, and astrocytes). Stem cells that are able to differentiate into cell types beyond those of the tissues in which they normally reside are said to exhibit **plasticity.** When a stem cell is found to give rise to multiple tissue types associated with different organs, the stem cell is referred to as **multipotent.**[1]

---

[1]The word "pluripotent" is sometimes used to describe stem cells that can differentiate into a *very wide range* of tissue types. In this report the term multipotent encompasses this type of stem cell.

Embryonic stem cells (ESCs) are derived from an early-stage embryo. Fertilization of an ovum by a sperm results in a zygote, the earliest embryonic stage (Figure 1). The zygote begins to divide about 30 hours after fertilization and by the third-to-fourth day, the embryo is a compact ball of 12 or more cells known as the morula. Five-to-six days after fertilization, and after several more cycles of cell division, the morula cells begin to specialize, forming a hollow sphere of cells, called a blastocyst, which is about 150 microns in diameter (one-seventh of a millimeter). The outer layer of the blasotocyst is called the trophoblast, and the cluster of cells inside the sphere is called the inner cell mass. At this stage, there are about 70 trophoblast cells and about 30 cells in the inner cell mass. The cells of the inner cell mass are multipotent stem cells that give rise to all cell types of the major tissue layers (ectoderm, mesoderm, and endoderm) of the embryo. In the past 3 years, it has become possible to remove these stem cells from the blastocyst and maintain them in an undifferentiated state in cell culture lines in the laboratory (NIH, 2001) (Figure 2). To be useful for producing medical therapies, cultured ESCs will need to be differentiated into appropriate tissues for transplantation into patients. Researchers are just beginning to learn how to achieve this differentiation.

Fetal stem cells are primitive cell types in the fetus that eventually develop into the various organs of the body, but research with fetal tissue so far has been limited to only a few cell types: **neural stem cells**, including **neural crest cells**; **hematopoietic stem cells**; and **pancreatic islet progenitors**. Neural stem cells, which are numerous in the fetal brain, can be isolated and grown in an undifferentiated form in culture, and they have been shown to differentiate into the three main types of brain cells (Brustle et al., 1998; Villa et al., 2000). These cells have been used in rodent models of Parkinson's disease (Sawamoto et al., 2001; Studer et al., 1998). Neural crest cells arise from the neural tube and migrate from it throughout the developing fetus. They are able to develop into multiple cell types, including the nerves that innervate the heart and the gut, non-neural cells of hormone-secreting glands, pig-

FIGURE 1

Stages of Development of the
Human Embryo

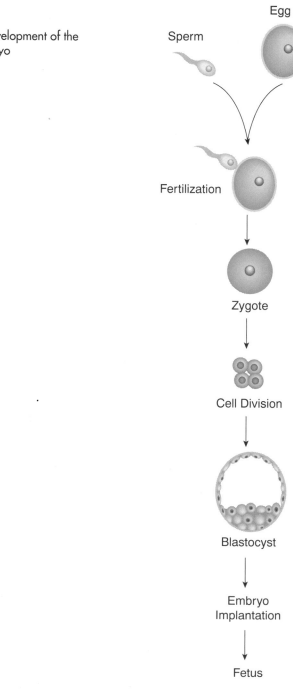

Egg

Sperm

Fertilization

Zygote

Cell Division

Blastocyst

Embryo
Implantation

Fetus

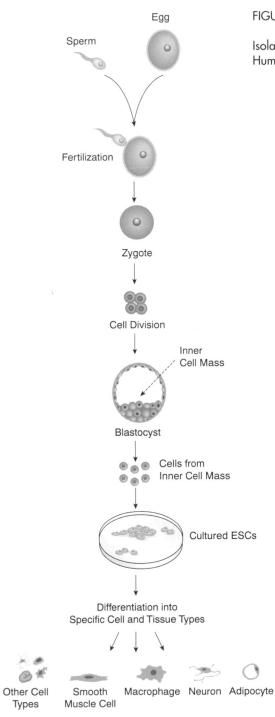

FIGURE 2

Isolation and Culture of
Human ESCs from Blastocysts

ment cells of the skin, cartilage and bone in the face and skull, and connective tissue in many parts of the body. Neural crest cells from mice have been cultured in the laboratory.

The fetal liver and blood are rich sources of hematopoietic stem cells, which are responsible for generating multiple cell types in blood, but their properties have not been extensively investigated. Although not part of the fetus, the umbilical cord and placenta are also rich sources of hematopoietic stem cells. Tissue extracted from the fetal pancreas has been shown to stimulate insulin production when transplanted into diabetic mice, but it is not clear whether this is due to a true stem cell, a more mature progenitor cell, or to the presence of fully mature insulin-producing pancreatic islet cells themselves (Beattie et al., 1997). Finally, multipotent cells called primordial germ cells have been isolated from the gonadal ridge, a structure that arises at an early stage of the fetus that will eventually develop into eggs or sperm in the adult. Germ cells can be cultured in vivo and have been shown to give rise to multiple cell types of the three embryonic tissue layers (Shamblott et al., 1998).

Adult stem cells are undifferentiated cells that occur in a differenti-ated tissue, such as bone marrow or the brain, in the adult body. They can renew themselves in the body, making identical copies of themselves for the lifetime of the organism, or become specialized to yield the cell types of the tissue of origin. Sources of adult stem cells include bone marrow, blood, the eye, brain, skeletal muscle, dental pulp, liver, skin, the lining of the gastrointestinal tract, and pancreas. Studies suggest that at least some adult stem cells are multipotent. For example, it has been reported that stem cells from the bone marrow, a mesodermal tissue, can give rise to the three major types of brain cells, which are ectodermal derivatives (Mezey et al., 2000) and that stem cells from the brain can differentiate into blood cells and muscle tissue (Bjornson et al., 1999), but these findings require verification. It is not clear whether investiga-tors are seeing adult stem cells that truly have plasticity or whether some tissues contain several types of stem cells that each give rise to only a few derivative types. Adult stem cells are rare, difficult to identify and purify,

and, when grown in culture, are difficult to maintain in the undifferentiated state. It is because of those limitations that even stem cells from
bone marrow, the type most studied, are not available in sufficient
numbers to support many potential applications of regenerative medicine. Finding ways to culture adult stems cells outside the body is a high
priority of stem cell research.

Additional terms used throughout this report are defined in the
Glossary. Although stem cells from all sources are important, the focus
of this report is on the characteristics and therapeutic potential of ESCs
and adult stem cells that have been at the center of scientific debate.

CHAPTER TWO

Adult Stem Cells

HEMATOPOIETIC STEM CELLS

The hope that many diseases can someday be treated with stem cell therapy is inspired by the historical success of bone marrow transplants in increasing the survival of patients with leukemia and other cancers, inherited blood disorders, and diseases of the immune system (Thomas and Blume, 1999). Nearly 40 years ago, the cell type responsible for those successes was identified as the hematopoietic stem cell (Till and McCullough, 1961). The ability of hematopoietic stem cells (HSCs) to self-renew continuously in the marrow and to differentiate into the full complement of cell types found in blood qualifies them as the premier adult stem cells (Figure 3).

HSCs are among the few stem cells to be isolated in adult humans. They reside in the bone marrow and under some conditions migrate to other tissues through the blood. HSCs are also normally found in the fetal liver and spleen and in umbilical cord and placenta blood.

There is a growing body of evidence that HSCs are plastic—that, at least under some circumstances, they are able to participate in the generation of tissues other than those of the blood system. A few studies have shown that HSCs can give rise to liver cells (Lagasse et al., 2000; Taniguchi et al., 1996; Thiese et al., 2001). Those findings have scientists speculating about the biological response of HSCs to disease or tissue damage and about the early differentiation of the

FIGURE 3

Blood Cell Differentiation from Hematopoietic Stem Cells (HSCs). HSCs normally divide to generate either more HSCs (self-renewal) or progenitor cells, which are precursors to various blood cell types. HSCs are found mainly in bone marrow, although T cells develop in thymus, and some other cell types develop from blood monocytes. Once HSCs partly differentiate into progenitor cells, further differentiation into one or a few types of blood cell is irreversible. Solid lines indicate known pathways; dash lines indicate pathways about which there is uncertainty.

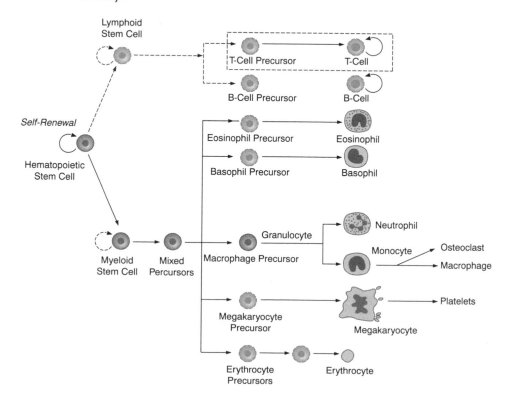

embryonic tissues into discrete layers. It was unexpected that a compo-
nent of blood could cross over a developmental separation to form a
tissue type that ordinarily has a completely different embryonic origin
(Lagasse et al., 2000). The findings noted above and other reports of
cardiac and muscle tissue formation after bone marrow transplantation in
mice (Bittner et al., 1999; Orlic et al., 2001) and of the development of
neuron-like cells from bone marrow (Brazelton et al., 2000; Mezey, et
al., 2000) have raised expectations that HSCs will eventually be shown to
be able to give rise to multiple cell types from all three germ layers. One
study has, in fact, demonstrated that a single HSC transplanted into an
irradiated mouse generated not only blood components (from the
mesoderm layer of the embryo), but also epithelial cells in the lungs, gut,
(endoderm layer) and skin (ectoderm layer) (Krause et al., 2001). If
HSCs are truly multipotent, their potential for life-saving regenerative
therapies may be considerably expanded in the future.

The full potential of bone marrow transplantation to restore a
healthy blood system in every needy patient is currently limited by the
unavailability of HSCs in the quantity and purity that are crucial for
successful transplantation. Because of their relative rarity (one in every
10,000 bone marrow cells) and the difficulty of separating them from
other components of the blood, so-called bone marrow stem cell transplants are generally impure (NIH, 2001). The significance of such
impurity is great. All cells of the body express on their surface a set of
molecules called histocompatibility (i.e. tissue compatibility) antigens. If
a patient receives a transplant of HSC cells from a donor that has histocompatibility antigens different from his own, the patient's body will
recognize and react to the cells as foreign. To increase the likelihood
that histocompatibility antigens will match, it is preferred that donors be
a related sibling of the transplant recipient. Even if their histocompatibility antigens do match, however, HSC transplants can be contami-
nated by T cells from the donor's immune system.

That contamination can cause the recipient's body to reject the
material or can produce an immune reaction in which the T cells of the
transplant attack the tissues of the recipient's body, leading to a poten-

tially lethal condition known as graft versus host disease. Although autologous transplants, in which material from a person is implanted into the same person (for example, when a cancer patient stockpiles his own blood in advance of chemotherapy or irradiation) solve the problem of immune system rejection, the inability to purify the material leads to the risk that diseased or cancerous cells in the transplant will later be reintroduced to the patient along with the stem cells.

In contrast, transplants of highly purified and concentrated populations of HSCs in mice have been shown to greatly reduce the incidence of graft versus host disease (Shizuru et al., 1996; Uchida et al., 1998). Purified and concentrated populations of autologous HSCs transplanted in breast cancer patients after chemotherapy have been shown to engraft more swiftly and with fewer complications (Negrin et al., 2000). Transplants of concentrated HSCs also have been shown to repopulate the blood more readily, reducing the period during which an individual is vulnerable to infection.

There is also evidence that transplants derived from umbilical cord blood are less likely to provoke graft versus host disease, possibly because the cells in cord blood are immature and less reactive immunologically (Laughlin, 2001). The quantity of HSCs present in cord blood and its attached placenta is small, and transplants from cord blood take longer to graft, but for children, whose smaller bodies require fewer HSCs, cord blood transplants are valuable, especially when there is no related sibling to donate HSCs (Gluckman et al, 2001). Banks of frozen umbilical cord and placenta blood (drawn out of the umbilical vein of the cord) are an important source of HSCs because the histocompatibility markers on the cells in these tissues can be identified and catalogued in advance of the need for a transplant.

Irving Weissman, who presented research findings on HSC transplantations at the workshop, has explored ways to improve the identification and purification of HSCs by looking for proteins on the surface of the stem cells that can be closely associated only with HSCs. Finding the specific profile of proteins that identifies HSCs, particularly those called long-term HSCs, is important, because these cells are believed to hold the key to future HSC therapies. Obtaining purified HSCs is a major challenge, and purification in a clinical setting is expensive and difficult.

Another major barrier to progress in HSC research and transplantation therapy is that it has not been possible to culture HSCs in vitro (outside the body), although recent studies of mouse HSCs grown in combination with components of the bone marrow have offered some preliminary promise (Ema et al., 2000; Moore et al. 1997). This stubborn and not insignificant obstacle is faced by researchers with all types of adult stem cells. If it were possible to expand the numbers of stem cells by growing them in culture or to stimulate their expansion in vivo (in the living body), the prospects for patients in need of stem cell transplants would be significantly improved. However, as Ernest Beutler pointed out at the workshop, finding a way to get HSCs to proliferate is not enough. In the long run, it is necessary to understand not only what activates HSCs to self-renew, but also what controls their decisions to differentiate into the various components of the blood and prevents them from developing into leukemic cells (Saito et al., 2000).

OTHER ADULT STEM CELLS

During the past 2 years, scientific reports of stem cells in other organs of adult mice—including brain, muscle, skin, digestive system, cornea, retina, liver, and pancreas—have cast a new light on the body's own capability to replenish its tissues (NIH, 2001). Their discovery has also fostered speculation that these cells exist in the adult human, that they have the characteristic of plasticity that enables them to change into precursors of cell types of other tissues, and that they will someday be used to produce the tissues for therapeutic use. The finding of stem cells in adult tissues, not all of which have been confirmed, offers a first glimpse at potential solutions to long-standing puzzles about why some human organs have a greater capacity for self-repair than others.

The idea of employing adult stem cells in certain therapeutic applications is appealing for several reasons. First, adult stem cells are naturally poised to generate a particular tissue, which might consist of several cell types, so they should be able to give rise to all the components of that tissue when transplanted into a patient. Second, some stem cells are able to migrate to injured tissue or other discrete sites in the body; for example, neural stem cells will migrate to tumor sites in the brain of a rodent (Aboody et al., 2000). This might provide more flexibility in choosing where to transplant them and more predictability in where they will localize after transplantation. Third, some adult stem cells are known to secrete growth factors that mobilize or protect other cells residing in the tissue that could increase the salutary effects of the transplant (Noble, 2000). It also might be possible to genetically engineer adult stem cells to produce higher levels of compounds normally produced in the body, to compensate for some deficiency in a patient's own tissues. Similarly, the cells could be engineered to secrete a therapeutic agent, such as a drug.

In other situations, the use of adult stem cells would be inappropriate—for example, the isolation and autologous transplantation of a person's stem cells suffering from a genetic disorder—in that case, the stem cells would carry the same incorrect genetic information. Transplantation of stem cells from a donor into another person will be subject to the problems of immune rejection, and this could be a substantial obstacle in time-critical situations, for example, spinal cord trauma or stroke, because characterizing the patient's tissues and finding a match in a short period of time will be difficult.

However, because recent findings of adult stem cells are so new and studies of them raise so many questions, even the most preliminary generalizations and conclusions as to therapeutic potential are tentative. As was noted by James Thomson at the workshop, the hematopoietic stem cell is the most characterized cell in the body, and "The amount of knowledge we have on other adult stem cells goes down dramatically from there."

First, human adult stem cells are rare and it is difficult to isolate a unique group of stem cells in pure form. So it is not surprising that what at first appears to be plasticity in a single adult stem cell type could be the result of a mixture of cells of different types, including different types of stem cells. At the workshop, Margaret Goodell explained how her research initially suggested that, given the right environment, stem cells from mouse muscle could be shown to produce not only muscle, but also components of blood. Later it became clear on rigorous testing that her sample contained two entirely different kinds of stem cells: one that formed blood and one that formed muscle. As she noted, the fact that the two types of stem cells were found in muscle might have interesting therapeutic uses, but in any case, it has not been demonstrated that a single type of stem cell in muscle exhibits this degree of plasticity.

An issue raised at the workshop was the need for more experiments that can show an unequivocal relationship between a stem cell and the tissues that are claimed to have arisen from it. In such an experiment, (an example of which is Krause's work on HSCs mentioned earlier in this chapter) a single, isolated stem cell would be chemically treated so that it incorporates a chemical "label" that will be passed on to all the cells that arise from it. If the labeled stem cell is injected into a mouse, any cell or tissue that is eventually found to have the label can be assumed to have come from the original single stem cell, and this is the kind of evidence for a definitive relationship that stem cell researchers are seeking.

A second factor that complicates adult stem cell research is that the environment in which stem cells grow or are placed to grow has an important but poorly understood effect on their fate—a theme that was echoed by many speakers at the workshop. For example, Iqbal Ahmad discussed prospects for retinal regeneration, which occurs naturally in goldfish but not in humans. Ahmad has isolated precursor cells in the mammalian eye that can be grown in culture for short periods and will develop into cells that appear to be retinal photoreceptors. If precursor cells from a mouse eye are transplanted into a normal mouse retina, they are not incorporated. In contrast, when transplanted into a diseased retina, the precursors begin to develop into photoreceptor-like cells and integrate into the tissue. Ahmad has not yet determined whether the integrated cells function normally.

What signals does the diseased retina provide that the normal retina does not? The cellular environment has important implications for how cells behave when they grow in a living organism (in vivo) and for what happens to them in culture (in vitro). For example, Ron McKay described how precursor cells taken from the mouse midbrain can be cultured in vitro to generate cells that appear to be dopaminergic neurons, but only for a very short time. (Dopaminergic neurons produce the chemical mediator L-dopa and are depleted in patients with Parkinson's disease.) Recalling that it is well known that the spinal cord generates motor neurons in response to signals that come from other tissues, McKay suggested that the capability of the precursors of the midbrain to make dopaminergic neurons might be transient in culture because they require stimulation from signals present only in the brain.

A third problem in understanding the capabilities of adult stem cells is the relationship of the cellular environment to the concept of plasticity in adult stem cells. Markus Grompe showed that HSCs and pancreatic stem cells can give rise to liver cells called hepatocytes that will repopulate a diseased mouse liver, demonstrating the plasticity of adult stem cells. However, in his experiment, HSCs and pancreatic stem cells were very inefficient in repopulating the liver relative to the ability of transplanted hepatocytes themselves. This could mean that the plasticity of adult stem cells is a marginal capacity that can be exploited only with a much greater understanding of the environmental signals that influence adult stem cells. No one knows what steps HSCs or pancreatic cells go through in generating a hepatocyte, or what signals cause them to migrate to the liver in the first place. Moreover, some of the apparent plasticity in adult stem cells is difficult to interpret because it has been accomplished in abnormal environments, for example, in mice that are immunologically impaired (Mezey et al., 2000) or sublethally irradiated (Brazelton et al., 2000).

Fourth, a major weakness of stem cell research asserted by Grompe is that most studies inadequately demonstrate that stem cells have produced a functionally useful cell in the organ. Most studies showing the plasticity of stem cells rely on the detection of proteins in the newly generated tissues that are commonly associated with a particular type of differentiated cell. But there is no consensus in the scientific community that the detection of a particular protein constitutes sufficient evidence that the cells and tissues formed are, in fact, fully functional and normal. Olle Lindvall, who works with Parkinson's disease patients, noted that in some experiments in which dopaminergic neurons generated in culture were grafted into the brain of an animal, it was not at all clear that the new neurons were fully functional. The relationship between stem cell type and environmental cues makes problematic the assumption that stem cells cultured in vitro can be expected to perform with predictable results when transplanted in vivo (Morrison, 2001). It might be possible someday to provide cues to reprogram one cell type into another and even to culture these cells in vitro, but evidence of the normal physiological and restorative function of adult stem cells is very limited today.

A fifth limitation relevant to immediate development of therapies based on adult stem cells is the inability to maintain these cells in culture for very long before they differentiate into their mature progeny. One can envision two therapeutic approaches to stem cells. In the first, stem cells themselves are implanted in a diseased or injured organ in the hope that they will give rise to the mature cells needed by that organ. In the second, the stem cells are stimulated to differentiate into the needed mature tissue outside the body, and that tissue is implanted in the organ. That adult stem cells are difficult to isolate, purify, and culture causes problems for either approach, although even the ability to culture stem cells for a limited time, including in the presence of other cells, could have therapeutic potential. An example is the use of autologous skin grafts for burn patients, in which healthy skin (which contains skin stem cells) is removed from the patient, cultured briefly outside the body, and grafted onto the patient's injured tissue. The grafts are not able to regenerate hair follicles and sweat glands, but are otherwise able to function normally. However, with a few exceptions, the appropriate culture conditions to sustain most adult stem cells indefinitely have yet to be found.

Very few stem cells, strictly defined, have even been isolated from adult human organs, in part because they constitute only a tiny fraction of the cells present and are not likely to be very distinct from the partially differentiated cells they give rise to as they mature and differentiate. For example, at the workshop, Fred Gage discussed his work with cadavers and brain biopsy material, wherein he found not stem cells, but rather what might be more mature types of neural cell precursors (Palmer et al., 2001). Those cells would differentiate into various neural tissues but would then stop dividing. Unlike stem cells, precursors and other subsequent intermediates generally undergo limited self-renewal in vivo and are committed to a pathway of differentiation into a specific tissue type. Researchers are, however, beginning to understand how a stem cell gives rise to precursor cells and, in at least one case, have used this information to manipulate that process. Using biochemical signals found in the cellular environment, rodent precursor cells in vitro were caused to revert into more primitive, multipotent stem cells (Kondo and Raff, 2000). The ability to "reprogram" a cell may be exploited someday to therapeutic ends; however, the reversal of the normal pathway of differentiation may have biological consequences not yet detected. Rigorous experimentation will be needed to evaluate the implications of this basic research finding for regenerative medicine.

Finally, the implications of what is known about *human adult* stem cells are often overlooked amid reports of successes with experiments in rodents that simulate heart attack, retinal disease, and diabetes. Confirmed reports of truly multipotent human adult stem cells are scarce. For its recently released report on stem cells, the National Institutes of Health could find few published accounts of the isolation of multipotent adult stem cells from human tissues (NIH, 2001). The much-publicized recent report of stem cells from human fat that produced cartilage, bone, and muscle (Zuk et al., 2001), for example, did not conclusively establish that the cells capable of performing this feat were fat cells. The authors of the paper conceded that the observation might have been due to the presence of another cell type, such as an HSC that had circulated out of blood and into fat. Without conclusive identification, the existence of a multipotent fat cell remains unconfirmed.

That there is little evidence of a wide array of human adult stem cells that can differentiate into multiple tissue types does not mean that they will not eventually be found, nor should it be interpreted to mean that the results of experiments with stem cells in rodents are not useful. Those results have prompted new theories about the source and significance of regenerative capabilities in all cells, about the process of cellular differentiation, and about the role of the physiological environment in inducing cells of all kinds to express their different characteristics (Blau et al., 2001).

All somatic cells in an organism contain the same genetic information, but it is not yet known what causes parts of the genetic code to be expressed in some cells and different parts to be expressed in others. This raises important and interesting questions about the ability of a cell of one type to become another type. Emphasizing how little is understood about the process that controls a cell's commitment to one course of action or another, Ihor Lemischka explained his findings that many genes found to be active in stem cells do not correspond to any known gene function ever described. A comparison of mouse and human HSCs shows that only about half of the genes expressed in the mouse HSCs correspond to genes expressed in human HSCs, so there are going to be differences as we move from experiments with mice to regenerative therapies in humans. Even the genetic programs that control the differentiation of human fetal liver stem cells and human HSCs, both of which evolve into the components of the blood, seem to be very different (Phillips et al., 2000). We need to understand much more about the differences between mouse and human stem cells if we are to harness their potential.

CHAPTER THREE

Embryonic Stem Cells

Embryonic stem cells (ESCs) are found in the inner cell mass of the human blastocyst, an early stage of the developing embryo lasting from the 4th to 7th day after fertilization. In normal embryonic development, they disappear after the 7th day, and begin to form the three embryonic tissue layers. ESCs extracted from the inner cell mass during the blastocyst stage, however, can be cultured in the laboratory and under the right conditions will proliferate indefinitely. ESCs growing in this undifferentiated state retain the potential to differentiate into cells of all three embryonic tissue layers. Research involving human ESCs is at the center of the ethical debate about stem cell use and potential in regenerative medicine. Embryos from which ESCs are extracted are destroyed in the process.

Several scientific questions are important when considering the potential of stem cells for use in regenerative medicine and the policy and ethical issues that arise:

• What properties of ESCs have promise for regenerative medicine?

• What direct evidence supports ESCs' effective use in regenerative medicine?

• What obstacles and risks are associated with the use of ESCs in regenerative medicine?

PROPERTIES OF ESCs IMPORTANT FOR REGENERATIVE MEDICINE

Human ESCs were successfully grown in the laboratory for the first time in 1998 (Thompson et al., 1998). Under appropriate culture conditions, ESCs have demonstrated a remarkable ability to self-renew continuously, that is, to produce more cells like themselves that are multipotent. As indicated at the workshop by Thomas Okarma and Ron McKay, ESC lines established from single cells have been demonstrated to proliferate through 300-400 population-doubling cycles. Human ESCs that have been propagated for more than 2 years also demonstrate a stable and normal complement of chromosomes, in contrast to the unstable and abnormal complement of embryonic cancer cell lines used in the past to study early stages of embryonic development. Careful monitoring of the aging ESC lines will be needed to evaluate the significance of genetic changes that are expected to occur over time.

Because human ESCs have only recently become available for research, most of what is known about ESCs comes from studies in the mouse, which, as noted in Chapter 2, cannot be presumed to provide definitive evidence of the capabilities of human cells.

Nevertheless, ESCs derived from mouse blastocysts have been studied for 2 decades and provide a critical baseline of knowledge about the biology and cultivation of these cells (Torres, 1998; Wobus and Boheler, 1999). The factors that permit the mouse ESC to continue replicating in the laboratory without differentiation and methods to trigger differentiation into different cell types that exhibit normal function have been actively explored. Among the types of cells derived from cultured mouse ESCs are fat cells, various brain and nervous system cells, insulin-producing cells of the pancreas, bone cells, hematopoietic cells, yolk sac, endothelial cells, primitive endodermal cells, and smooth and striated muscle cells, including cardiomyocytes—heart muscle cells (Odorico et al., 2001).

Experience with mouse ESCs has provided clues to methods for culturing human ESCs and leading them to differentiate. Mouse ESCs will proliferate in an undifferentiated state in the presence of a biochemical called leukemia inhibitory factor (LIF), but the culture conditions required to keep human ESCs from differentiating include growing them in petri dishes on a layer of mouse embryonic fibroblasts (referred to as "feeder cells") in a medium containing serum from cows. The feeder cells are inactivated, so they are not dividing and expanding, but they produce growth factors that sustain the ESCs. The mechanism of how feeder cells maintain the proliferation of undifferentiated ESCs is unknown. Such in vitro culturing presents certain theoretical hazards to the use of stem cells for regenerative medicine, such as the spread of viruses and other infectious agents not normally found in humans. When removed from feeder cells and grown in suspension (in liquid), human ESCs form aggregated balls of cells called "embryonic bodies," which have been reported to give rise to a multiplicity of cell types representing all three layers of embryonic tissue development (Itskovitz-Eldor et al., 2000; Reubinoff et al., 2000; Schuldiner et al., 2000). Evidence of the differentiation in culture includes detection of the products of genes associated with different cell types and in some cases by the characteristic shapes that are peculiar to different cell types. Cells derived from human embryonic bodies include "rhythmically contracting cardiomyocytes, pigmented and nonpigmented epithelial cells, and neural cells displaying an exuberant outgrowth of axons and dendrites" (Odorico et al., 2001). In other experiments, cells arising from human ESCs have been reported to express genes associated with liver and pancreas function (Schuldiner et al., 2000). Human ESCs grown in coculture with mouse bone marrow stromal cells have been reported to produce colonies of human hematopoietic precursors and ultimately cells from the blood (Kaufman et al., 1999).

Further evidence of the multipotent capability of human ESCs is based on studies in an in vivo setting. Human ESCs injected into mice form a type of benign tumor called a teratoma that is made up of tissues from all three embryonic layers. The tissues that arise in the tumor are often advanced, organized, and complex, and include teeth, gut, hair follicles, skin, epithelium, muscle, bone, cartilage, lung tissue, and neural cells (Thompson et al., 1998). The experiments showed the capability of ESCs to produce a variety of tissues, but the results also highlight the complexity of the biological "program" of tissue development that can unfold in different biological environments. These results also emphasize the abnormal, potentially neoplastic potential of ESCs when placed into unnatural environments.

Major questions remain about the genetic or environmental factors in the body that control the fate of ESCs and about the importance of different factors during various stages of cell differentiation. Even on the basis of the limited findings, however, the ability to grow human ESCs in vitro and to have them differentiate in the laboratory makes them an important and unique tool with which to conduct the basic research that is critical for the foundation of future regenerative therapies. It has been possible, for example, to create a lineage of mouse ESCs that generate neural cell precursors (Li et al., 1998). Studies of the genes turned on and off as cells begin to differentiate, which are already under way with ESCs, will permit a better understanding of the genetic controls important in tissue differentiation (Duncan et al., 1998). In vitro studies of ESCs also provide an opportunity to explore the role of biochemicals produced in the normal cellular environment that induce stem cells to differentiate, to migrate to a site needing repair, and to assimilate into tissues (Schuldiner et al., 2000).

EVIDENCE SUPPORTING THE POTENTIAL OF ESCs FOR USE IN REGENERATIVE MEDICINE

At the workshop, James Thomson and Thomas Okarma suggested that human ESCs will someday provide a potentially unlimited source of cells, differentiated in vitro, for transplantation therapies involving the liver, nervous system, and pancreas. Irving Weissman alluded to the possible use of ESCs to enhance the success of whole-organ transplantation. If HSCs derived from human ESCs could be successfully transplanted into the blood system of a transplant recipient (by using immunosuppressive drugs), any further implant tissue (say kidney or pancreas) developed with the same ESCs would not, in theory, be rejected by the recipient because the immune cells produced in the recipient's blood by the HSCs would see the implant tissue as "self".

But that is a long way off, as Marcus Grompe noted, in as much as no one has yet demonstrated any in vivo reconstitution of an organ's function in either humans or experimental animals with cells derived from human ESCs. Moreover, ESCs in tissue culture give rise to a mixture of cell types all at once, and biochemical, tissue-culture, and molecular-biology techniques to control and limit differentiation require much further investigation.

Because human ESCs have only recently become available for research, and because public funding for such research has been limited, studies of how well ESCs or their differentiated tissues perform physiologic functions has been largely conducted with mouse models. Ron McKay described progress made in coaxing the in vitro differentiation of human ESCs into insulin-producing cells that might be useful in treating diabetes, but he also noted that studies have already been conducted with analogous mouse cells transplanted into mice that have diabetes and that partial restoration of insulin regulation was observed (Lumelsky et al., 2001). Other studies have demonstrated that mouse ESCs can be successfully transplanted into rodents that have Parkinson's disease symptoms and partially relieve these symptoms (Studer et al., 1998). Similarly, studies suggest that mouse ESCs can be transplanted into animals that have spinal-cord injuries and partially restore neural function (McDonald et al., 1999).

Those studies provide promise, but not definitive evidence, that similar treatments could be effective in humans. Human ESCs will need to be tested in primate models, such as those for Parkinson's disease and diabetes mellitus in the rhesus monkey. Methods for transplanting ESCs need to be developed, as do means of establishing whether the cells develop and function properly after transplantation. In some cases, it will be important to ensure that the transplanted cells or tissues are incorporated and positioned properly relative to existing tissues, such as in heart and neural tissue; the three-dimensional, cell-to-cell interactions will play important roles in the functioning of an organ. Other cells, like pancreatic islet cells, or hematopoietic cells, will require less complex incorporation.

Also, the large-scale propagation of human ESCs in culture will require that they can be grown without feeder cells (Odorico et al., 2001). Research is needed to elucidate the mechanisms of feeder cells in repressing differentiation and to find alternatives to them, at the same time eliminating the potential that an animal virus from the feeder cells might be transferred to the ESCs.

Finally, it was noted earlier that the chromosomes of human ESCs have been shown to be stable in tissue culture. This does not mean however, that ESC lines will not be subject to the random mutations that affect all cell lines as they age. In cells from humans and other animals, approximately one mutation occurs every time a cell divides. A cell that has divided 200 times in culture therefore can be expected to harbor approximately 200 different mutations (Kunkel and Bebeneck, 2000). So far, there have been no studies published about the changes that may have occurred in existing stem cell lines. Vigilant monitoring of the integrity of existing cell lines is essential to allow understanding of the impact of long-term culture, and new stem cell lines may need to be developed in the future.

Obstacles and Risks Associated with the Use of ESCs

In addition to demonstrating the functional effectiveness of ESC transplants, it is necessary to identify and minimize, or eliminate, the risks that ESCs might pose. Two identifiable risks are tumor formation and immune rejection. As noted earlier, human ESCs injected into mice can produce a benign tumor made up of diverse tissues; this response is believed to be related to the multipotency of the undifferentiated cells in an in vivo environment. However, in a small number of short-term studies in mice, human ESCs that have been allowed to begin the process of differentiation before transplantation have not resulted in

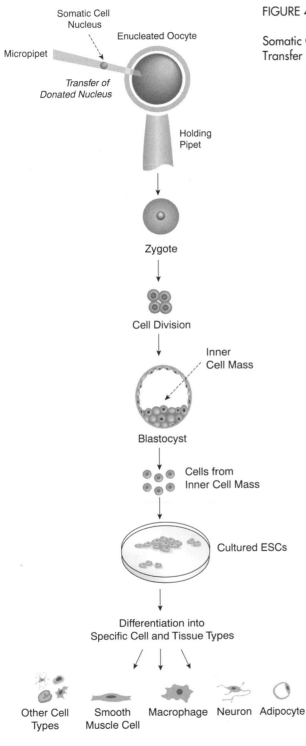

FIGURE 4

Somatic Cell Nuclear
Transfer (SCNT)

significant tumor formation (Odorico et al., 2001). Obviously, this is a critical problem to understand and control.

It is too early to tell, therefore, whether it will be appropriate to use human ESCs directly in regenerative medicine. A great deal obviously must be elucidated about how the body controls the differentiation of stem cells, and this has yet to be reliably reproduced in vitro. Also, the behavior of ESCs implanted in a specific organ has not been well studied. It might someday be possible to add growth factors with a transplant to stimulate the production of a particular cell type or multiple cell types. "Inducer tissues" that interact with stem cells might be co-transplanted with ESCs to achieve a similar result. Those possibilities are still in experimental investigation.

In another respect, the possible problems associated with ESC transplantation are common to all transplantation, such as the risk of infection and the risk of tissue rejection. As discussed in Chapter 2, rejection is a serious obstacle to successful transplantation of stem cells and tissues derived from them. It has been suggested that ESCs provoke less of an immune reaction than a whole-organ transplant, but it is unclear whether that will be true of the regenerated tissues derived from ESCs. Some types of cells (such as dendritic cells, immune system cells, and vascular endothelial cells) carry more of the histocompatibility antigens that provoke immune reactions than other cells. Those types are present in the tissues of whole organs; they connect an organ with the bloodstream and nervous system. However, tissue derived in vitro from ESCs, such as liver tissue, would not contain such cells and therefore would theoretically trigger a milder immune response; this assumes that techniques for controlling differentiation of ESCs will be available. In addition, the liver cells likely would not be devoid of all surface antigens, and so, in the absence of other techniques to reduce transplant rejection, the use of immunosuppressive drugs will still have to be used, with attendant risks of infection and toxicity.

Although difficult to conceive, the creation of a very large number of ESC lines might be one way to obtain a diversity of cells that could theoretically increase the chances of matching the histocompatibility antigens of a transplant recipient. It has also been suggested that ESCs could be made less reactive by using genetic engineering to eliminate or introduce the presence of surface antigens on them (Odorico, 2001). An exact genetic match between a transplant recipient and tissue generated from ESCs could also, in theory, be achieved by using somatic cell nuclear transfer to create histocompatible ESCs (Figure 4). Cells created with this technique would overcome the problem of immune rejection. However, it might to not be appropriate to transplant such cells in a person with a genetically based disease, since the cells would carry the same genetic information. In any case, an understanding of how to prevent rejection of transplanted cells is fundamental to their becoming useful for regenerative medicine and represents one of the greatest challenges for research in this field.

CHAPTER FOUR

Opportunities for and Barriers to Progress in Stem Cell Research for Regenerative Medicine

KEY SCIENTIFIC QUESTIONS

Both adult and embryonic stem cells can contribute to the development of regenerative medicine. Embryonic stem cells (ESCs) have the advantage of multipotency and have shown themselves to be readily culturable in the laboratory. Although the degree of plasticity of adult stem cells is still unknown and there are difficulties in purifying and culturing them, the only proven stem cell-based medical therapies that are currently available rely on adult-derived stem cells from bone marrow and skin, and adult stem cells from other tissues might someday provide therapies that stimulate the body's own regenerative potential. Because of a misunderstanding of the state of knowledge, there may be an unwarranted impression that widespread clinical application of new therapies is certain and imminent. In fact, stem cell research is in its infancy, and there are substantial gaps in knowledge that pose obstacles to the realization of new therapies from either adult or embryo-derived stem cells.

Bone marrow transplantation is a case in which clinical application proceeded without a thorough understanding of the underlying biology, but the success of the technique has improved dramatically as the understanding has grown (Thomas et al., 1999). We might not need a universal understanding of the origins and embryonic development of stem cells, but we do need to know the answers to some fundamental questions:

- What causes stem cells to maintain themselves in an undifferentiated state?
- What cues do cells use to tell them when to start or stop dividing?
- What genetic and environmental signals affect differentiation?
- What physiological properties guide the functional integration of newly generated tissues into existing organs?

The scientific investigations that will answer those questions need to be comprehensive and repeated before researchers can make strong claims about the capabilities of stem cells. Because stem cell research is relatively new, it is important to build a scientific foundation that can support the research community's ability to evaluate and confirm new findings and demonstrations. The pillars of this foundation were identified at the workshop. They include markers that characterize specific types of stem cells; markers that distinguish stages of a stem cell's commitment to differentiate into a particular cell lineage; profiles of gene expression in stem cells and their progeny; standard procedures for isolating stem cells from the body; techniques to propagate them reliably; and consensus on the physiological or other criteria that confirm restoration of tissue function following stem cell transplantation.

As knowledge of stem cells grows, investigators will be able to ask meaningful questions about therapeutic approaches, including whether to implant cells in an undifferentiated state or a differentiated state, and which of the various sources of stem cells are best suited to address a specific clinical need. It may become apparent that combined therapies (transplanting multiple stem cell types, or using gene therapy in combination with stem cells transplants) will be needed, depending on such factors as the stage of a particular disease or the age of a patient. For now, all of these questions must wait for the establishment of a more firm scientific foundation.

Because observing the behavior of tissue in vivo is difficult and the results can be confounded in many ways, sources of human stem cells that can be cultured in vitro are perhaps the most critical need of investigators. They will permit many more questions to be posed and answered. Many more experiments can be completed with cultured cells in the same amount of time and with the same degree of effort as in living organisms. Moreover, data from in vitro studies allow more insightful and better-defined experiments to be developed in living organisms. Access to ESCs is likely to ultimately determine the rate at which scientists make progress in this field. In fact, the successful cultivation of postnatal and adult sources of stem cells for regenerative medicine is likely to advance more rapidly if the study of ESCs proceeds and cells from different sources can be compared. ESCs exhibit many properties whose improved understanding could assist researchers in modifying adult stem cells to achieve better growth in culture and greater capacity for controlled differentiation.

USE OF EMBRYOS FOR STEM CELL RESEARCH IS CONTROVERSIAL

A second major obstacle to the development of new medical therapies based on stem cells is opposition to ESC research on ethical, moral, or religious grounds. No field of biological science has been more controversial than that involving human reproduction. Contraception, abortion, and in vitro fertilization have all provoked major debate and controversy in this country and abroad. Stem cell research also touches on some of the most fundamental issues with which society has grappled over the centuries, including the definition of human life and the moral and legal status of the human embryo.

The workshop provided an opportunity for the committee to hear both from those who support ESC research and from those who oppose it on ethical grounds. The various speakers on the workshop panel articulated the main ethical arguments, which are summarized below. The committee acknowledges the importance and value of a dialogue that respects the many differing perspectives. Although it is not within our charge to judge the validity of the ethical arguments for and against this research, we believe it is appropriate to address aspects of the debate that touch upon scientific questions about the biology and derivation of stem cells.

The most basic objection to ESC research is rooted in the fact that such research deprives a human embryo of any further potential to develop into a complete human being. For those who believe that the life of a human being begins at the moment of conception, ESC research violates tenets that prohibit the destruction of human life and the treatment of human life as a means to some other end, no matter how noble that end might be.

There are widely divergent views on this subject. For example, in testimony to the National Bioethics Advisory Commission, Rabbis Elliot Dorff and Moshe Dovid Tendler explained that in Jewish law and tradition the embryo has no moral status until 40 days after implantation. Until it is born, the child is viewed as a part of its mother's body, and its own life is believed to begin only when the child is born. Eggs and sperm mixed together in a petri dish have no legal status, because they are not even part of a human being unless implanted in a woman's womb. In the same forum, Abdulaziz Sachedina discussed the Muslim tradition, which accords legal and moral status to the fetus only after ensoulment takes place, at the end of the fourth month of pregnancy. Because in both of those belief systems there is a mandate to save human life wherever possible, human ESC research can be deemed acceptable if it is conducted reasonably and ethically (National Bioethics Advisory Commission, 1999).

In past Roman Catholic tradition, the Aristotelian view that life begins 40 days after conception was adopted by Augustine of Hippo and Thomas Aquinas and was maintained by the church for centuries. In 1869, however, a supplanting view that we cannot know with certainty when human life begins became established (Noonan, 1970). This view, which is currently held by the Catholic Church, requires that human life be protected at the earliest possible time, which is taken to be at conception. Protestant denominations hold diverse views: some conservative

Protestants reject the use of embryos for research, but most accept ESC research. Moreover, not everyone who rejects embryonic stem cell research is either religious or conservative. Every federal commission (e.g., National Bioethics Advisory Commission, 1999) that has addressed research on human embryos and fetuses has, in light of the many differing perspectives, called for respect for these entities as forms of human life.

For those who hold the views that human life begins at conception and that the moral obligation to preserve human life outweighs any potential health benefits of ESC research for regenerative medicine, the only morally acceptable position would be to adopt a complete prohibition on human ESC research without regard to the method of embryo production or whether the research is publicly or privately funded.

Views that require less than a complete prohibition, however, permit consideration of trade-offs in defining what is acceptable. Many of the other positions rely on distinctions made about the source of the ESCs for research. Thus, one viewpoint would allow the use of embryonic cells already in laboratory culture but would prohibit the destruction of additional embryos to derive new cell lines. Another would permit the derivation and use of new cell lines as long as the cells originated in "excess" embryos that were produced by in vitro fertilization for reproductive purposes but are no longer needed for such purposes. Still another would permit the use of cells derived from embryos created specifically for the research from eggs and sperm donated by volunteers who are unrelated to each other and have no reproductive intent. The potential for creating embryos with the somatic cell nuclear transfer (SCNT) technique represents yet another approach that does not even involve fertilization of an egg by a sperm. Each of those constructs can pose its own ethical dilemmas. LeRoy Walters summarized many of the dilemmas and the diverse public-policy responses that have been adopted by various countries (see Box). As Kevin FitzGerald pointed out at the workshop, this issue is complex and confusing and poses challenges not only to science, but to society.

International Perspective on Public Policy on Human ESC Research

Germany:
Prohibits the derivation and use of human ESCs from blastocysts.

United States:
As articulated by President Bush on August 9, 2001, permits federal funding only for research using cells from approximately 60 stem cell lines identified by the National Institutes of Health as having been derived from excess human embryos prior to the August 9 announcement. There is currently no federal law or policy prohibiting the private sector from creating stem cells by in vitro fertilization or by the SCNT technique for the purpose of research, but as this report went to print, legislative prohibitions were under consideration. The policies of most individual states also currently permit private funding of the use of human ESCs derived from excess in vitro fertilization embryos, embryos created by in vitro fertilization for the purpose of research, and embryos created with the SCNT technique, although a few states have banned some of these.

France:
Permits the use of human ESCs and their derivation from superfluous embryos not needed by the genetic parents for reproduction. (This approach has also been recommended by ethical advisory committees in Canada, Japan, and Germany.)

United Kingdom:
Permits the use of human ESCs and their derivation from leftover or superfluous embryos not needed by the genetic parents for reproduction, from embryos created for research purposes by in vitro fertilization, and embryos created with the SCNT technique. (The option of allowing human creation of ESCs for research purposes with the SCNT technique is also being considered in Italy, France, Australia, Israel, and Holland.)

It is not for this committee to comment on the validity of the ethical or moral arguments for or against any of the alternatives. Indeed, it is highly likely that even the members of the committee would differ in what is acceptable to them personally. It is, however, appropriate for the committee to reiterate a few key points to increase the focus and clarity of the various ethical debates.

First, arguments in favor of imposing constraints or even an outright prohibition on ESC research are frequently supported by the assertion that research on stem cells from adult tissues alone will lead to the development of the sought-after medical therapies. In his presentation at the workshop, for example, David Prentice cited many reports as supporting the argument that research on adult stem cells has all the necessary scientific potential and represents a morally less problematic alternative that obviates the need for research on ESCs. But Prentice also pointed out that much of this evidence is suggestive rather than definitive and that the hurdles so far encountered in research on adult stem cells suggest that predictions of success are highly speculative. As discussed in Chapter 2, the evidence indicates that there are substantial potential problems in realizing this goal. Stem cells in adult mammalian tissues are rare and difficult to isolate, and very few stem cell types have been confirmed to exist in adult human tissues. Most types of adult stem cells are difficult to grow in culture, and their potential plasticity has not been clearly established. Much of the work that is used to support the argument that adult stem cells can substitute for ESCs was done only in mice or other animal models, which might or might not prove applicable to humans (Chen et al., 2001; Clarke et al., 2000; Jackson et al., 2001; Kocher et al., 2001; Krause et al., 2001; Orlic et al., 2001; Ramiya et al., 2000; Torrente et al., 2001; Wang et al., 2000), or reported work performed with human hematopoietic stem cells (Bhardwaj et al., 2001; Cashman and Eaves, 2000; Colter et al., 2000; Gilmore et al., 2000; Laughlin et al., 2001), which is not generalizable to other cell types. It should also be noted that the study of human ESCs is likely to advance some applications of adult stem cells in the future.

Second, the creation of stem cells with the SCNT technique is problematic to some because the technique is similar to that used for reproductive cloning. There is a scientific rationale for not foreclosing this avenue of research and for distinguishing clearly between SCNT to prevent transplant rejection and SCNT to create a fetus. Theoretically, the SCNT technique could produce genetically identical stem cells that could give rise to tissues that would not be rejected by a transplant recipient's immune system. That is an attractive option because such a histocompatible transplant would not prompt the types of medically serious and potentially life-threatening immunological responses encountered by transplants of tissue from foreign donors.

Third, the smaller the number of cell lines in use, the lower the genetic diversity that they represent. A prohibition on the derivation of new cell lines might result in research that focuses on cell lines that are not optimal and might preclude the replacement of inferior materials with more efficient cell lines. Experience with other kinds of cells in culture has shown that cell lines can be expected to accumulate mutations that reduce their suitability and safety for research (Kunkel and Bebenek, 2000). There is little evidence that ESC lines will behave any differently.

Fourth, it has been suggested that it is biologically preferable to derive stem cells from embryos created specifically for research rather than from surplus embryos at in vitro fertilization clinics, although both employ similar techniques in the initial stages. Several ideas underlie that suggestion. Embryos from couples who have turned to in vitro fertilization because of infertility might have inherent, but as yet unrecognized, biological defects. From a broader genetic perspective, couples who seek treatment for infertility might not be representative of the genetic diversity of society as a whole. In addition, it might be preferable to obtain embryos that have not been frozen before stem cells are derived from them, inasmuch as freezing could have unexpected effects. Each of these concerns has only a theoretical basis, and there is currently little evidence with which to evaluate the relative merits of stem cells created specifically for research versus those derived from surplus embryos.

PUBLIC FUNDING PROVIDES THE BEST OPPORTUNITIES FOR THERAPEUTIC ADVANCES

Given the many unanswered questions about the biology of stem cells, the successful development of new medical therapies depends in large part on the performance of an enormous amount of basic research. Basic research is defined as systematic study directed toward greater knowledge or understanding of the fundamental aspects of phenomena and of observable facts without specific applications, processes, or products in mind. Since World War II, basic research has been the traditional domain of public funding, which optimizes opportunities for scientific advance in several ways:

• The differing roles of public and private investment in research.
• The increased likelihood of advancing knowledge through a broad spectrum of diverse research activities.
• The increased accessibility of research results as a consequence of public funding.
• The enhanced opportunities for oversight and regulation associated with public funding.

The Roles of Public and Private Resources for Basic Research

The National Institutes of Health (NIH) is the largest federal sponsor of health research, with a budget of more than $20 billion in FY2001. Even in 1997, pharmaceutical and biotechnology firms exceeded that total for overall biomedical research (Institute of Medicine, 1998), spending about $19 billion and $8 billion respectively in that year. In 2001, an estimated $50 billion will be spent on US biomedical research by public and private funding sources combined (Nathan et al., 2001). But within the larger system, NIH is the primary sponsor of the basic biomedical research that produces new fundamental knowledge. By its own accounting, NIH estimated that 62% of its budget was devoted to basic research in FY1996. By comparison, basic research represented an average of only about 14 percent of all private-sector pharmaceutical R&D in the 1990s, with pharmaceuticals representing the major area of concentration for private R&D according to a recently completed National Research Council report on trends in support of research (NRC, 2001). Although not-for-profit private entities, such as the Howard Hughes Medical Institute, also support basic research, private-sector efforts are dominated by for-profit companies that focus their research investments on product-related applications, such as new drugs, diagnostic tools, and medical devices that cure, detect, or prevent disease.

Arti Rai, an expert in the legal aspects of biotechnology and health care whose work addresses the interactions between the public and private sectors in biomedical research, spoke at the workshop on this issue. She stated, "Because basic research is often far removed from commercial application, it is unlikely to be pursued at the levels at which it should be pursued by private companies that need to satisfy shareholders with short-term commercial results." Absent public funding, she said, even fiscally conservative economists tend to agree that socially optimal levels of basic research will not be pursued. She noted, however, that Geron, which provided the funding for the stem cell discoveries of James Thomson and John Gearhart, had stepped into the breach when public funding for embryonic stem cell research was unavailable.

Importance of Multiple Avenues of Research

A prohibition on federal funding of ESC research would limit progress not only by limiting funds, but also by limiting the number of scientists who participate in the research. Dr. Thomas O'Karma, President and Chief Executive Officer of Geron Corporation, which funded and holds licenses to the stem cell discoveries of James Thomson and John Gearhart, commented at the workshop that it is frustrating not to be able to distribute these cells more widely to NIH-funded investigators for them to extend and validate the data Geron is generating.

Although in principle academic scientists could accept private funding to pursue research that is subject to federal restrictions, this may not be a viable option for many. NIH can revoke a scientist's funding for violating federally imposed restrictions. If a federally funded research institution were to support an individual scientist in such a violation, public funding of the institution as a whole could also be threatened. But drawing a sufficiently clear line between activities and infrastructure supported by the federal government and those supported only by the private sector in a single laboratory or university can be difficult. The establishment of separate privately supported laboratories that are free of federal funds, such as the University of Wisconsin's WiCell Institute, entails substantial costs to duplicate infrastructure, equipment, and personnel (Gulbrandsen, 2001), and such measures may not be feasible for many academic institutions.

Another issue is that confining the research effort to a small number of entities may diminish the rate of discovery and knowledge development. As discussed at the workshop by Arti Rai, the history of scientific innovation strongly indicates that basic research and its applications are best developed by multiple entities pursuing a variety of research questions. She gave examples from the automobile, aircraft, radio, and semiconductor industries, which went through a stage of development during which progress was slow in large part due to the fact that many of the key technologies were held exclusively by individual companies and not widely accessible. It was only when the companies agreed to share their interdependent technologies that progress accelerated. In general, during periods of dominance by a single entity with monopoly control over crucial patents, scientific and technological development can be impeded.

In contrast, public funding of basic biomedical research has historically resulted in the results of the research being widely available to other scientists. Publicly funded researchers typically publish their research results in scientific journals, and this mechanism for information exchange can stimulate progress. (See also next section.) Even patenting of publicly funded research need not be a deterrent to progress if such patented research is licensed with terms that enable broad dissemination of the patented research. A notable example is the Cohen-Boyer patent on the recombinant-DNA technique, which emerged from public funding and was held by the University of California, San Francisco, and by Stanford University. Those institutions licensed the research widely at reasonable rates, and many analysts attribute the successful evolution of recombinant-DNA technology to those licensing arrangements. Arti Rai believes that the traditional academic focus on the importance of wide dissemination of fundamental knowledge has encouraged universities to shy away from exclusive licensing of the most fundamental research. Although never exercised, the Bayh-Dole Act contains a provision that gives the federal government a limited legal right to compel licensing. Comparable authorities to compel-licensing of privately funded basic research results are more limited and depend on specific legal authorities, such as a finding of an antitrust violation.

Need for Accessibility

A strong feature of publicly funded basic biomedical research in the United States is the widespread dissemination of experimental methods and findings through scientific publication as they emerge. Although the lines between industry and academe are increasingly blurred, the academic norm of free information exchange generally persists. Many benefits emerge from open access to data and methods. As scientists stay abreast of findings in a field, they are better able to refine their own research agendas, permitting an informed and broad base of research activities, which is important for innovation. Peer review strengthens the rigor of research, in that the design of experiments and reporting of data in grant proposals and publications must meet accepted scientific standards. Access to data also permits replication of a study, which is critical for authenticating scientific findings. It is important to note that many of the stem cell papers published to date, although heavily publicized by the mass media, have not yet passed the essential test of replication and scientific confirmation and must therefore be considered less than conclusive.

Research Oversight

The federal government in general and NIH in particular exert tremendous influence on the research that they fund through the mechanism by which they approve studies, the priorities they set, policy-making from informed-consent procedures to patent-seeking, and making the results of their research investments publicly available. When heightened public scrutiny is warranted, NIH can implement even more rigorous review and oversight mechanisms, as was the case for the controversial research involving recombinant DNA (see Box). Other means for regulating research exist, such as the passage of federal and state laws, but the public funding mechanism is the major means by which NIH influences the type of research performed and the way it is conducted. Public funding would guarantee regulatory oversight for stem cell research, allowing, for example, a careful informed consent procedure for obtaining ESCs that is subject to public scrutiny.

Peer review, as noted above, helps to ensure the quality of research proposals. As part of peer review, the importance of the research questions addressed and the methods used to answer them are considered by leading scientists with appropriate expertise. In some fields, review committees also use the input of experts in ethics and representatives of the public who are stakeholders in the research, ensuring greater public

The Recombinant DNA Advisory Committee

The Recombinant DNA Advisory Committee, or RAC, was established by the director of NIH in 1975. Its creation was the result of concern among scientists and the public about the safety of laboratory studies aimed at introducing new DNA into organisms. The committee, after addressing laboratory safety and commercial development of recombinant DNA techniques and release of altered organisms into the environment, established benchmarks for review and approval of protocols for applying the techniques of gene transfer to humans. Both technical and ethical issues were considered. RAC advises the NIH director as to whether specific research proposals should be approved and gives guidance on recombinant-DNA research and relevant ethical issues.

Scientists and physicians make up the majority of RAC's membership with lawyers, social scientists, ethicists, and stakeholders from the public. Because it was a federal advisory committee, its meetings were announced in the *Federal Register* and were open to the public. When important new scientific projects came before the committee for review, mass-media attention would often be intense, giving the group's recommendations extensive coverage.

Although RAC officially was limited to providing advice to the NIH director on whether studies should be approved, its power extended beyond NIH-sponsored research. In a recent rechartering of RAC, nonvoting representatives from various other federal agencies were included. The Food and Drug Administration and Environmental Protection Agency indicated that any products developed using recombinant DNA must comply with RAC guidance.

Source: Institute of Medicine, *Society's Choices: Social and Ethical Decision Making in Biomedicine.* 1995.

accountability. If specific rules govern a field of research, such as the need for informed consent of research volunteers or the requirements for research subjects of both sexes or various ethnic groups, the review process considers whether the requirements have been met or, if not, whether sufficient justification is provided for deviating from them. In short, public funding engenders considerable opportunity for shaping the types of research that are approved. In general, privately funded investigators are subject to less oversight and review, although activities such as the pursuit of patents or marketing approvals from the Food and Drug Administration represent other mechanisms for oversight that are less relevant to basic research.

CHAPTER FIVE

Findings and Recommendations

S tem cell research offers unprecedented opportunities for developing new treatments for debilitating diseases for which there are few or no cures. Stem cells also present a new way to explore fundamental questions of biology, such as determining the basic mechanisms of tissue development and specialization, which will be required for the development of therapies. However, our society holds diverse views about the morality of using early embryos for research, and we find ourselves searching for a consensus on how to proceed with this new avenue of research. Provocative and conflicting claims about the biology and biomedical potential of adult and embryonic stem cells have been made both inside and outside the scientific community. The committee considered those claims in light of the meaning and importance of the preliminary data from recent stem cell experiments. The following findings and recommendations constitute the final result of the committee's deliberations on these issues.

Finding 1: Experiments in mice and other nonhuman animals are necessary but not sufficient for medical advances in human regenerative medicine. There are substantial biological differences between animal and human development and between animal and human stem cells, although the full range of similarities and differences is not understood.

Recommendation: Studies with *human* stem cells are

essential to make progress in the development of treatments for *human disease*, and this research should continue.

Finding 2: Current scientific data indicate that there are important biological differences between adult and embryonic stem cells and among adult stem cells found in different types of tissue. The therapeutic implications of these biological differences are not clear, and additional scientific data are needed on all stem cell types. Adult stem cells from bone marrow have so far provided most of the examples of successful therapies for replacement of diseased or destroyed cells. Their potential for fully differentiating into other cell types (such as brain, nerve, and pancreas cells) is still poorly understood and remains to be clarified. In contrast, embryonic stem cells studied in animals clearly are capable of developing into multiple tissue types and capable of long-term self-renewal in culture, features that have not yet been demonstrated with many adult stem cells. The application of stem cell research to therapy for human disease will require much more knowledge about the biological properties of all types of stem cells. The best available scientific and medical evidence indicates that research on both embryonic and adult human stem cells will be needed. Moreover, research on embryonic stem cells will be important to inform research on adult stem cells, and vice versa.

Recommendation: Although stem cell research is on the cutting edge of biological science today, it is still in its infancy. Studies of both embryonic and adult human stem cells will be required to most efficiently advance the scientific and therapeutic potential of regenerative medicine. Research on both adult and embryonic human stem cells should be pursued.

Finding 3: Over time, all cell lines in tissue culture change, typically accumulating harmful genetic mutations. There is no reason to expect stem cell lines to behave differently. In addition, most existing stem cell lines have been cultured in the presence of nonhuman cells or serum that could lead to potential human health risks. Consequently, vigilant monitoring of the integrity of existing cell lines is essential. In addition, the generation of new stem cell lines is likely to be important to replace those that become inviable and to increase understanding of the impact of long-term cell culture.

Recommendation: While there is much that can be learned using existing stem cell lines if they are made widely available for research, concerns about changing genetic and biological properties of these stem cell lines necessitate continued monitoring as well as the development of new stem cell lines in the future.

Finding 4: High-quality, publicly funded research is the wellspring of medical breakthroughs. Although private, for-profit research plays a critical role in translating the fruits of basic research into medical advances that are broadly available to the public, the status of stem cell research is far from the point of providing therapeutic products. Without public funding of basic research on stem cells, progress toward medical therapies is likely to be hindered. In addition, public funding offers greater opportunities for regulatory oversight and public scrutiny of stem cell research.

Recommendation: Human stem cell research that is publicly funded and conducted under established standards of open scientific exchange, peer-review, and public oversight offers the most efficient and responsible means to fulfill the promise of stem cells to meet the need for regenerative medical therapies.

Finding 5: Conflicting ethical perspectives surround the use of embryonic stem cells in medical research, particularly where the moral and legal status of human embryos is concerned. The differing perspectives are difficult to reconcile. Given the controversial nature of research with fetal and embryonic tissues, restrictions and guidelines for ethical conduct of such research have been developed.

Recommendation: If the federal government chooses to fund human stem cell research, proposals to work on human embryonic stem cells should be required to justify the decision on scientific

grounds and should be strictly scrutinized for compliance with existing and future federally mandated ethical guidelines.

Finding 6: The use of embryonic stem cells is not the first scientific advance to raise public concerns about ethical and social issues in bio-medical research. Recombinant-DNA techniques likewise raised questions and were subject to intense debate and public scrutiny. In that case, a national advisory body, the Recombinant DNA Advisory Committee, was established at the National Institutes of Health to ensure that the research met with the highest scientific and ethical standards.

Recommendation: A national advisory group composed of outstanding researchers, ethicists, and other stakeholders should be established at NIH to oversee research on human embryonic stem cells. The group should include leading experts in the most current scientific knowledge relevant to stem cell research who can evaluate the technical merit of any proposed research on human embryonic stem cells. Other roles for the group could include evaluation of potential risks to research subjects and ensuring compliance with all legal requirements and ethical standards.

Finding 7: Regenerative medicine is likely to involve the implantation of new tissue in patients with damaged or diseased organs. A substantial obstacle to the success of transplantation of any cells, including stem cells and their derivatives, is the immune-mediated rejection of foreign tissue by the recipient's body. In current stem cell transplantation procedures with bone marrow and blood, success hinges on obtaining a close match between donor and recipient tissues and on the use of immunosuppressive drugs, which often have severe and potentially life-threatening side effects. To ensure that stem cell-based therapies can be broadly applicable for many conditions and people, new means of overcoming the problem of tissue rejection must be found. Although ethically controversial, the somatic cell nuclear transfer technique promises to have that advantage. Other options for this purpose include genetic manipulation of the stem cells and the development of a very large bank of ES cell lines.

Recommendation: In conjunction with research on stem cell biology and the development of potential stem cell therapies, research on approaches that prevent immune rejection of stem cells and stem cell-derived tissues should be actively pursued. These scientific efforts include the use of a number of techniques to manipulate the genetic makeup of stem cells, including somatic cell nuclear transfer.

References

Aboody, K.S., Brown, A., Rainov, N.G., Bower, K.A., Liu, Shaoxiong, Yang, W., Small, J.E., Herrlinger, U., Ourednik, V., Black, P.M., Breakefield, X.O., and Snyder, E.Y. (2000). Neural stem cells display extensive tropism for pathology in adult brain: evidence from intracranial gliomas. Proc Natl Acad Sci USA 97(23):12846-51.

Beattie, G.M., Otonkoski, T., Lopez, A.D., and Hayek, A. (1997). Functional beta-cell mass after transplantation of human fetal pancreatic cells: differentiation or proliferation? Diabetes 46:244-8.

Bhardwaj, G., Murdoch, B., Wu, D., Baker, D.P., Williams, K.P., Chadwick, K., Ling, L.E., Karanu, F.N., Bhatia, M. (2001). Sonic hedgehog induces the proliferation of primitive human hematopoietic cells via BMP regulation. Nat Immunol 2:172-80.

Bittner, R.E., Schofer, C., Weipoltshammer, K., Ivanova, S., Streubel, B., Hauser, E., Freilinger, M., Hoger, H., Elbe-Burger, A., and Wacthler, F. (1999). Recruitment of bone-marrow-derived cells by skeletal and cardiac muscle in adult dystrophic mdx mice. Anat Embryol (Berl) 199:391-6.

Bjornson, C.R., Rietze, R.L., Reynolds, B.A., Magli, M.C., and Vecovi, A.L. (1999). Turning brain into blood: a hematopoietic fate adopted by adult neural stem cells in vivo. Science 283:534-7.

Blau, H.M., Brazelton, T.R., and Weimann, J.M. (2001). The evolving concept of a stem cell: entity or function? Cell 105:829-41.

Brazelton, T.R., Rossi, F.M., Keshet, G.I., and Blau, H.M. (2000). From marrow to brain: expression of neuronal phenotypes in adult mice. Science 290:1775-9.

Brustle, O., Choudhary, K., Karram, K., Huttner, A., Murray, K., Dubois, D.M., and McKay, R.D. (1998). Chimeric brains generated by intraventricular transplantation of human brain cells into embryonic rats. Nat Biotechnol 16 (11):1040-4.

Cashman, J.D. and Eaves, C.J. (2000). High marrow seeding efficiency of human lymphomyeloid repopulating cells in irradiated NOD/SCID mice. Blood 96:3979-81.

Chen J., Li, Y., Wang, L., Zhang, Z., Lu, D., Lu, M., and Chopp, M. (2001) Therapeutic benefit of intravenous administration of bone marrow stromal cells after cerebral ischemia in rats. Stroke 32:1005-1.

Clarke, D.L., Johansson, C.B., Wilbertz, J., Veress, B., Nilsson, E., Karlstrom, H., Lendahl, U., and Frisen, J. (2000). Generalized potential of adult neural stem cells. Science 288:1660-3.

Colter, D.C., Class, R., DiGirolamo, C.M., and Prockop, D.J. (2000). Rapid expansion of recycling stem cells in cultures of plastic-adherent cells from human bone marrow. Proc Natl Acad Sci USA 97:3213-8.

Duncan, S.A., Navas, M.A., Dufor, D., Rossant, J., and Stoffel, M. (1998). Regulation of a transcription factor network required for differentiation and metabolism. Science 281: 692-5.

Ema, H., Takano, H., Sudo, K., and Nakauchi, H. (2000). In vitro self-renewal division of hematopoietic stem cells. J Exp Med 192:1281-8.

Ferrari, G., Cusella-DeAngelis, G., Coletta, M., Paolucci, E., Stornaiuolo, A., Cossu, G., and Mavilio, F. (1998). Muscle regeneration by bone marrow-derived myogenic progenitors. Science 279:1528-30.

Gilmore, G.L., DePasquale, D.K., Lister, J., and Shadduck, R.K. (2000). Ex vivo expansion of human umbilical cord blood and peripheral blood CD34(+) hematopoietic stem cells. Exp Hematol 28:1297-1305.

Gluckman, E., Rocha, V., and Chevret, S. (2001). Results of unrelated umbilical cord blood hematopoietic stem cell transplant. Transfus Clin Biol 8(3):146-54.

Gulbrandsen, C.E. August 1, 2001, Testimony to the US Senate Subcommittee on Labor, Health, Human Services, Education, and Related agencies, Committee on Appropriations. Washington, D.C.

Gussoni, E., Soneoka, Y., Stickland, C.D., Buzney, E.A., Khan, M.K., Fline, A.F., Kunkel, L.M., and Mulligan, R.C. (1999). Dystrophin expression in the mdx mouse restored by stem cell transplantation. Nature 401:390-4.

Institute of Medicine. (1995). Society's choices: social and ethical decision making in biomedicine. Washington, D.C.: National Academy Press.

Institute of Medicine. (1998). Scientific opportunities and public needs: improving priority setting and public input at the National Institutes of Health. Washington, D.C.: National Academy Press.

Itskovitz-Eldor, J., Schulding M., Karesenti, D., Eden., A., Yanuka, O., Amit, M., Soreq, H., and Benvenisty, N. (2000). Differentiation of human embryonic stem cells into embryoid bodies comprising the three embryonic germ layers. Mole Med 5:88-95.

Jackson K.A., Majka, S.M., Wang, H., Pocius, J., Hartley, C.J., Majesky, M.W., Entman, M.L., Michael, L.H., Hirschi, K.K., and Goodell, M.A. (2001). Regeneration of ischemic cardiac muscle and vascular endothelium by adult stem cells. J Clin Invest 107:1395-1402.

Kaufman, D.S., Lewis, R.L., Auerbach, R., et al. (1999). Directed differentiation of human embryonic stem cells into hematopoeitic colony forming cells. Blood 94 (supplement 1, part 1 of 2):34a.

Kocher, A.A., Schuster, M.D., Szabolcs, M.J., Takuma, S., Burkhoff, D., Wang, J., Homma, S., Edwards, N.M., and Itescu, S. (2001). Neovascularization of ischemic myocardium by human bone-marrow-derived angioblasts prevents cardiomyocyte apoptosis, reduces remodeling and improves cardiac function. Nat Med 7:430-6.

Kondo, T., and Raff, M. (2000). Oligodendrocyte precursor cells reprogrammed to become multipotential CNS stem cells. Science 289(5485):1754-7.

Krause, D.S., Theise, N.D., Collector, M.I., Henegariu, O., Hwang, S., Gardner, R., Neutzel, S., and Sharkis, S.J. (2001). Multi-organ, multi-lineage engraftment by a single bone marrow-derived stem cell. Cell 105:369-77.

Kunkel, T.A., and Bebenek, K. (2000). DNA replication fidelity. Annu Rev Biochem 69:497-529.

Lagasse, E., Connors, H., Al-Dhalimy, M., Reitsma, M., Dohse, M., Osborne, L., Wang, X., Finegold, M., Weissman, I.L., and Grompe, M. (2000). Purified hematopoietic stem cells can differentiate into hepatocytes in vivo. Nat Med 6:1229-34.

Laughlin, M.J. (2001). Umbilical cord blood for allogeneic transplantation in children and adults. Bone Marrow Transpl 27:1-6.

Laughlin, M.J., Barker, J., Bambach, B., Koc, O.N., Rizzieri, D.A., Wagner, J.E., Gerson, S.L., Lazarus, H.M., Cairo, M., Stevens, C.E., Rubinstein, P., and Kurtzberg, J. (2001). Hematopoietic engraftment and survival in adult recipients of umbilical-cord blood from unrelated donors. New Eng J Med 344:1815-22.

Li, M. Pevny, L., Lovell-Badge, R., and Snith, A. (1998). Generation of purified neural precursors from embryonic stem cells by lineage selection. Curr Biol 8:971-4.

Lumelsky, N., Blondel, O., Laeng, P., Velasco, I., Ravin, R., and McKay, R. (2001). Differentiation of embryonic stem cells to insulin-secreting structures similar to pancreatic islets. Science 292:1389-94.

McDonald, J.W., Liu, X.Z., Qu Y Liu, S., Mickey, S.K., Turetsky, D., Gottlieb,D.I., and Choi, D.W. (1999) Transplanted embryonic stem cells survive, differentiate and promote recovery in injured rat spinal cord. Nat Med 5:1410-2.

Mezey, E., Chandross, K.J., Harta, G., Maki, R.A., and McKercher, S.R. (2000). Turning blood into brain: cells bearing neuronal antigens generated in vivo from bone marrow. Science 290:1779-82.

Moore, K.A., Ema H., and Lemischka, I.R. (1997). In vitro maintenance of highly purified, transplantable hematopoietic stem cells. Blood 89:4337-47.

Morrison, S.J. (2001). Neuronal differentiation: proneural genes inhibit gliogenesis. Curr Bio 11:R349-51.

Nathan, D.G., Fontanarosa, P.B., and Wilson, J.D. (2001). Opportunities for medical research in the 21st century. JAMA 285(5):533-4.

National Bioethics Advisory Commission. (1999). Ethical issues in human stem cell research, volume III, Religious Perspectives. Rockville, Md.: National Bioethics Advisory Commission.

National Institutes of Health. (July 2001). Stem cells: scientific progress and future research directions, p 46. http://www.nih.gov/news/stemcell/scireport.htm.

National Research Council. (2001). Trends in federal support of research and graduate education. Washington, D.C.: National Academy Press.

Negrin, R.S., Atkinson, K., Leemhuis, T., Hanania, E., Juttner, C., Tierney, K., Hu, W.W., Johnston, L.J., Shizuru J.A., Stockerl-Goldstein, K.E., Blume, K.G., Weissman, I.L., Bower S., Baynes, R., Dansey, R., Karanes, C., Peters, W., and Klein, J. (2000). Transplantation of highly purified CD32+Thy-1+ hematopoietic stem cells in patients with metastatic breast cancer. Biol Blood Marrow Transpl 6:262-71.

Noble, M. (2000). Can neural stem cells be used to track down and destroy migratory brain tumor cells while also providing a means of repairing tumor-associated damage? Proc Natl Acad Sci USA 97(23):12393-5.

Noonan, J.T. (1970). An almost absolute value in history, pp. 1-59. In Noonan, J.T., editor, The morality of abortion — legal and historical perspectives. Cambridge, Mass.: Harvard University Press.

Odorico, J.S., Kaufman, D.S., and Thomson, J.A. (2001). Multilineage differentiation from human embryonic stem cell lines. Stem Cells 19:193-204.

Orlic D., Kajstura, J., Chimenti, S. Jakoniuk, I., Anderson, S.M., Li, B. Pickel, J., McKay, R., Nadal-Ginard, B., Bodine, D.M., Leri, A., and Anversa, P. (2001). Bone marrow cells regenerate infarcted myocardium. Nature 410:701-5.

Palmer, T.D., Schwartz, P.H., Taupin, P., Kaspar, B., Stein, S.A., and Gage, F.H. (2001). Cell culture: progenitor cells from human brain after death. Nature 411(6833):42-3.

Perry, D. (2000). Patients' voices: the powerful sound in the stem cell debate. Science 287:1423.

Phillips, R.L., Ernst, R.E., Brunk, B., Ivanova, N., Mahan, M.A., Deanehan, J.K. , Moore, K.A., Overton, G.C., and Lemischka, I.R. (2000). The genetic program of hematopoietic stems cells. Science 288(5471):1635-40.

Ramiya, V.K., Maraist, M., Arfors, K.E., Schatz, D.A., Peck, A.B., and Cornelius, J.G. (2000). Reversal of insulin-dependent diabetes using islets generated in vitro from pancreatic stem cells. Nat Med 6:278-82.

Reubinoff, B.E., Per, M.F., Fong, C.Y., Trounson, A., and Bongso, A. (2000). Embryonic stem cell lines from human blastocysts: somatic differentiation in vitro. [published erratum, Nat Biotechno 2000; 18:559.] Nat Biotechno 18:299-404.

Saito, Y., Uzuka, Y., Sakai, N., Suzuki, S., and Toyota, T. (2000). Bone marrow transplant 25(11):1209-11.

Sawamoto K, Nakao N, Kakishita K, Ogawa Y, Toyama Y, Yamamoto A, Yamaguchi M, Mori K, Goldman SA, Itakura T, and Okano H. (2001). Generation of dopaminergic neurons in the adult brain from mesencephalic precursor cells labeled with a nestin-GFP transgene. J Neurosci 21:3895-3903.

Schuldiner, M., Yanuka, O., Istkovitz-Eldor, J. Melton, D., and Benvenistry, N. (2000). Effects of eight growth factors on the differentiation of cells derived from human embryonic stem cells. Proc Natl Acad Sci USA 907:11307-12.

Shamblott, M.J., Axelman, J., Wang S., Bugg, E.M., Littlefield, J.W., Donovan, P.J., Blumenthal, P.D., Huggins, G.R., and Gearhart, J.D. (1998). Derivation of pluripotent stem cells from cultured human primordial germ cells. Proc Natl Acad Sci wS.A. 95:13726-31.

Shizuru, J.A., Jerabeck, L., Edwards, C.T., and Weissman, I.L. (1996). Biol Blood Marrow Transpl 2:3.

Studer, L., Tabar, V., and Mckay, R.D.G. (1998). Transplantation of expanded mesencephalic precursors leads to recovery in Parkinsonian rats. Nat Neurosci 1:290-5.

Taniguchi, H., Toyoshima, T., Fukau, K., and Nakauchi, H. (1996). Presence of hematopoietic stem cells in the adult liver. Nat Med 2:198-203.

Theise, N.D., Nimmakayalu, M., Gardner, R., Illei, P.B., Morgan, G., Teperman, L., Henegariu, O., and Krause, D.S. (2001). Liver from bone marrow in humans. Hepatology 32:11-16.

Thomas, E.D., and Blume, K.G. (1999). Historical markers in the development of allogeneic hematopoietic cell transplantation. Biol Blood Marrow Transpl 5:341-6.

Thomson, J.A., Itskovitz-Eldor, J., Shapiro, S.S., Waknitz, M.A., Swiergiel, J.J., Marshall, V.S., and Jones, J.M. (1998). Embryonic stem cell lines derived from human blastocysts. Science 282:1145-7.

Till, J.E., and McCullough, E.A. (1961). A direct measurement of the radiation sensitivity of normal mouse bone marrow cells. Radiat Res 14:213-22.

Torrente Y, Tremblay, J.P., Pisati, F., Belicchi, M., Rossi, B., Sironi, M., Fortunato, F., El Fahime, M., D'Angelo, M.G., Caron, N.J., Constantin, G., Paulin, D., Scarlato, G., and Bresolin, N. (2001). Intraarterial injection of muscle-derived CD34+Sca-1+ stem cells restores dystrophin in mdx mice. J Cell Biol 152:335-48.

Torres, M. (1998). The use of embryonic stem cells for the genetic manipulation of the mouse. Curr Top Dev Biol 36:99-114.

Uchida, N., Tsukamoto, A., He, D., Friera, A.M., Scollay, R., and Weissman, I.L. (1998). High doses of purified stem cells cause early hematopoietic recovery in syngeneic and allogeneic hosts. J Clin Invest 101:961-6.

Villa A., Snyder, E.Y., Vescovi, A., and Martinez-Serrano, A. (2000) Establishment and properties of a growth factor-dependent perpetual neural stem cell line from the human CNS. Exp Neurol 161(1):67-84.

Wang, J.S., Shum-Tim, D., Galipeau, J., Chedrawy, E., Eliopoulos, N., and Chiu Ray, C.J. (2000). Marrow stromal cells for cellular cardiomyoplasty: feasibility and potential clinical advantages. J Thorac Cardiovasc Surg 120:999-1006.

Wobus, A.M., and Boheler, K.R. (1999). Embryonic stem cells as a developmental model in vitro. Cells Tissues Organs 165:3-4.

Zuk, P.A., Zhu, M., Mizuno, H., Huang, J., Futrell, J.W., Katz, A.J., Benhaim, P., Lorenz, P., and Hedrick, M.H. (2001). Multilineage cells from human adipose tissue: implications for cell-based therapies. Tiss Eng 7(2):211-28.

Glossary

Adult stem cell - An undifferentiated cell that is found in differentiated adult tissue, can renew itself, and can (with certain limitations) differentiate to yield all the specialized cell types of the tissue from which it originated.

Antigen - Any substance, usually a protein, that stimulates an immune response.

Autologous transplant - Transplanted tissue that is derived from the intended recipient of the transplant. Such a transplant helps avoid complications of immune rejection.

Blastocyst - A preimplantation embryo of 30-150 cells and 4-7 days of age. The blastocyst consists of a sphere made up of an outer layer of cells (the trophectoderm), a fluid-filled cavity (the blastocoel), and a cluster of cells on the interior (the inner cell mass).

Bone marrow - The soft, living tissue that fills most bone cavities and contains hematopoietic stem cells from which all red and white blood cells evolve. The bone marrow also contains mesenchymal (stroma) stem cells that a number of cells types come from, including chondrocytes, which produce cartilage.

Bone marrow cell - Refers to both hematopoietic cells and mesenchymal (stromal) cells.

Bone marrow stem cell - Refering to one of at least two types of multipotent stem cells: hematopoietic stem cell and mesenchymal stem cell.

Chromosomes - Nucleic acid-protein structures in the nucleus of a cell. Chromosomes are composed chiefly of DNA, the carrier of hereditary information. Chromosomes contain genes, working subunits of DNA that carry the genetic code for specific proteins, interspersed with large amounts of DNA of unknown function. A normal human body cell contains 46 chromosomes; a normal human gamete (egg or sperm), 23 chromosomes.

Cytoplasm - The contents of a cell, other than the nucleus cytoplasm, consists of a fluid containing numerous structures, known as organelles, that carry out essential cell functions.

Dendrite - Extension of a nerve cell, typically branched and relatively short, that receives stimuli from other nerve cells.

Differentiation - The process whereby an unspecialized early embryonic cell acquires the features of a specialized cell such as a heart, liver, or muscle cell.

DNA - A chemical, deoxyribonucleic acid, found primarily in the nucleus of cells. DNA carries the instructions for making all the structures and materials the body needs to function.

Ectoderm - The upper, outermost of the three primitive germ layers of the embryo that will give rise to the skin, hair, nails, nerve, and brain including the retina of the eye.

Embryo - In humans, the developing organism from the time of fertilization until the end of the eighth week of gestation, when it becomes known as a fetus.

Embryonic germ cell - Cells found in a specific part of the embryo/fetus called the gonadal ridge, and normally develop into mature gametes.

Embryonic stem cell (ESC) - Primitive (undifferentiated) cell from the embryo that have the potential to become a wide variety of specialized cell types.

Endoderm - The lower, inner of the three primitive germ layers of the embryo that will give rise to the epithelial layers of the lungs and bronchi, pharynx, gastrointestinal tract, liver, pancreas, and urinary bladder.

Fertilization - The process whereby male and female gametes unite.

Gene - A functional unit of heredity that is a segment of DNA and located in a specific site on a chromosome. Genes generally direct the formation of an enzyme or other protein.

Genome - The complete genetic material of an organism.

Germ cell - A gamete, that is, a sperm or egg, or a cell that can become a sperm or egg. All other body cells are somatic cells.

Germ layers - The three initial tissue layers arising in the embryo—endoderm, mesoderm, and ectoderm—from which all other somatic tissue-types develop.

Gonadal ridge -Anatomic site in the early fetus where primordial germ cells are formed.

Graft-versus-host disease - A condition that occurs following bone marrow transplant in which the donor's immune cells, in the transplanted marrow, make antibodies against the host's tissues.

Hematopoietic stem cell (HSC) - A stem cell from which all red and white blood cells evolve.

Hepatic - Relating to the liver.

Histocompatible - The immunological characteristic of cells or tissue that causes them to be tolerated by another cell or tissue; that allows some tissues to be grafted effectively to others.

Immune system cells - White blood cells or leukocytes that originate from the bone marrow. They include antigen-presenting cells, such as dendritic cells, T and B lymphocytes, and neutrophils, among many others.

In vitro - From the Latin for, "in glass"; in a laboratory dish or test tube; an artificial environment.

In vitro fertilization (IVF) - An assisted reproduction technique in which fertilization is accomplished outside the body.

In vivo - In the living subject; a natural environment.

Inner cell mass - The cluster of cells inside the blastocyst. These cells give rise to the embryonic disk of the later embryo and ultimately the fetus. They are the source of embryonic stem cells.

Lipid - Any one of a group of fats or fat-like substances characterized by their insolubility in water and solubility in fat solvents such as alcohol, ether, and chloroform.

Lymphocyte - A type of white blood cell that is part of the body's cellular immune system; present in the blood and lymphatic tissue.

Macrophage - A lymphocyte that has left the circulation and settled and matured in a tissue. Because of their placement in the lymphoid tissues, macrophages serve as the major scavenger of the blood, clearing it of abnormal or old cells and cellular debris as well as pathogenic organisms.

Mesenchyme - Connective tissue arising from multiple germ layers consisting of unspecialized cells. A number of cell types come from the mesenchyme, including the cells that give rise to collagen, muscle, cartilage, and bone.

Mesoderm - The middle of the three primitive germ layers of the embryo. These cells occur between the ectoderm and endoderm and give rise to most of the cardiovascular system, blood cells and bone marrow, the skeleton, smooth and striated muscles, and parts of the reproductive and excretory system.

Morula - A solid mass of 12 or more cells that resembles a mulberry, occurring at 3 to 4 days after fertilization and that results from the cleavage of the zygote.

Multipotent - Capable of differentiating into multiple cell types associated with different organs.

Neoplastic - Having the characteristic of potentially malignant growth.

Neural stem cell - A stem cell that can give rise to the different types of cells of the nervous system. Neural stem cells are found in certain areas of the adult brain, in embryos, fetuses, newborns, and juveniles.

Neuron - The key data-processing cell of the nervous system. Each neuron has a cell body and one or more processes (extensions) called dendrites and axons. Neurons function by the initiation and conduction of electrical impulses that are transmitted to other neurons or cells.

Ovum - An egg, the female germ, or sex, cell produced in the ovaries.

Placenta - The oval or discoid spongy structure in the uterus from which the fetus derives its nourishment and oxygen

Plasticity - The ability of a cell to differentiate into a cell type beyond the tissue in which it normally resides.

Somatic cell - Any cell of a plant or animal other than a germ cell or germ-cell precursor.

Somatic cell nuclear transfer - The transfer of a cell nucleus from a somatic cell into an egg from which the nucleus has been removed.

Stem cell - A cell that has the ability to divide for indefinite periods in culture and to give rise to specialized cells.

Stromal cell - A non-blood cell that is derived from blood organs, such as bone marrow or fetal liver, which is capable of supporting growth of blood cells in vitro. Stromal cells that make up the matrix within the bone marrow are derived from the mesenchyme and give rise to fat and cartilage.

Surface antigen - Proteins on the surface of cells that are capable of detection by antibodies or other means. These may stimulate an immune response.

T cell - A type of white blood cell that is of crucial importance to the immune system. Immature T cells (termed T-stem cells) migrate to the thymus gland in the neck, where they differentiate into various types of mature T cells and become active in the immune system. T cells that are potentially activated against the body's own tissues are normally killed or changed ("down-regulated") during this maturation process.

Tissue culture - Growth of tissue in vitro on an artificial media for experimental research.

Undifferentiated - Not having changed to become a specialized cell type.

White blood cell - Also known as a leukocyte. These cells normally protect against infection by, for example, ingesting bacteria or secreting antibodies. White blood cells are formed from the undifferentiated stem cell that can give rise to all blood cells. Those in the bone marrow may become any of the five types of white blood cells. Those in the spleen and lymph nodes may become lymphocytes, or monocytes, and those in the thymus can become lymphocytes (T-lymphocytes).

Zygote - The cell formed by the union of male and female germ cells (sperm and egg, respectively).

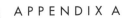

Committee Biographies

Bert Vogelstein, MD (*Chair*), Johns Hopkins Oncology
Center and Howard Hughes Medical Institute, is professor of
oncology and pathology at the Johns Hopkins University and
an investigator of the Howard Hughes Medical Institute. He
has broad expertise in the molecular and cell biological processes underlying human disease. His research seeks to
understand the complex sequence of genetic alterations that
are responsible for transforming a normal colon cell to a
malignant one. He has won numerous awards, including the
American Cancer Society Medal of Honor, the Gairdner
Foundation International Award in Science, the Baxter Award
from the Association of American Medical Colleges, the
Clowes Memorial Award from the American Association for
Cancer Research, the Pezcoller Award from the European
School of Oncology, the William Beaumont Prize in Gastro-
enterology from the American Gastroenterological Association, the Karnofsky Memorial Award from the American
Society for Clinical Oncology, the William Allan Award from
the American Society of Human Genetics, the Paul Ehrlich
and Ludwig Darmstaedter Prize from the Paul Ehrlich Foundation, the Richard Lounsbery Award from the National
Academy of Sciences, the Louisa Gross Horwitz Prize
from Columbia University, and the Charles S. Mott Prize
from the General Motors Cancer Research Foundation. Dr.
Vogelstein is a member of the American Academy of Arts and

Sciences and the National Academy of Sciences. He received his BA in mathematics from the University of Pennsylvania and his MD from the Johns Hopkins University School of Medicine. He performed his internship and residency in pediatrics at the Johns Hopkins Hospital.

Barry R. Bloom, PhD, Harvard School of Public Health, is dean of the faculty and professor of immunology and infectious diseases at the Harvard School of Public Health. His research interests include immunology, resistance to infectious disease, vaccine development, and international health. Dr. Bloom chairs the WHO UNAIDS Vaccine Advisory Committee and serves on the National AIDS Vaccine Research Committee. He recently received a major grant from the Bill and Melinda Gates Foundation for an AIDS-prevention initiative in Nigeria. He was a member of the National Advisory Council of the National Institute for Allergy and Infectious Diseases and the US National Vaccine Advisory Committee. He was president of the American Association of Immunologists and President of the Federation of American Societies for Experimental Biology. He serves on the Scientific Advisory Board of the National Center for Infectious Diseases of the Centers for Disease Control and Prevention, and the National Advisory Board of the Fogarty International Center at the National Institutes of Health. Dr. Bloom is chairman of the Board of Trustees of the International Vaccine Institute. He was cochair of the Board on Global Health of the Institute of Medicine. He received the first Bristol-Myers Squibb Award for Distinguished Research in Infectious Diseases, shared the Novartis Award in Immunology, and was the recipient of the Robert Koch Gold Medal for lifetime research in infectious diseases. Dr. Bloom is a member the Institute of Medicine, the American Academy of Arts and Sciences, and the National Academy of Sciences. He received his AB, and an honorary SD, from Amherst College, his MA from Harvard University; and his PhD from the Rockefeller University.

Corey Goodman, PhD, University of California, Berkeley, and Howard Hughes Medical Institute, is Evan Rauch Professor of Neuroscience in the Department of Molecular and Cell Biology and a Howard Hughes Medical Institute investigator. He is director of the Wills Neuroscience Institute, whose mission is to build bridges across traditional academic boundaries from genes and genomes to brain and behavior. His expertise is in developmental neurobiology—using genetic analysis to unravel the mechanisms that control the "wiring" of the brain. He was elected a member of the National Academy of Sciences in 1995 and in January 2001 became chair of the Board on Life Sciences of the National Research Council. He serves as president of the McKnight Endowment Fund for neuroscience. He is cofounder of two biotechnology companies—Exelixis and Renovis—and is cochair of the Renovis Scientific Advisory Board. He is the recipient of the Foundation IPSEN Neuronal Plasticity Prize, the J. Allyn Taylor International Prize in Medicine, the Gairdner Foundation International Award for Achievement in Medical Sciences, the Ameritec Foundation Prize, the Wakeman Award, and the March-of-Dimes Prize in Developmental Biology. He received his BS in biology from Stanford University and his PhD in developmental neurobiology from the University of California, Berkeley.

Patricia King, JD, Georgetown University Law Center, is Carmack Waterhouse Professor of Law, Medicine, Ethics and Public Policy at the Georgetown University Law Center. Her research interests include law, bioethics, and public policy. She has concentrated on reproductive and scientific issues related to embryos and fetuses. She is a member of the Institute of Medicine and a fellow of both the Hastings Center and the Kennedy Institute of Ethics. She is a member of the Institute of Medicine Committee on Assessing the System for Protecting Human Subjects, the cochair for policy of the National Institutes of Health (NIH) Embryo Research Panel, and a member of Working Group to Advise ACD (NIH) on Guidelines and Oversight of stem cell research. She has previously served on the Institute's Committee on Organ Procurement and Transplantation Policy and Committee on Perinatal Transmission of HIV. She received her JD from Harvard Law School.

Guy McKhann, MD, Johns Hopkins University School of Medicine, is professor of neurology, Zanvyl Krieger Mind/Brain Institute of the Johns Hopkins University School of Medicine. His research involves neurologic and cognitive outcomes of cardiac surgery and definition of a new form of Guillain-Barre syndrome. He is a member of the Institute of Medicine (IOM) and has served on the IOM Committee to Plan the Symposium on Neuroscience and Brain Research. He is also a member of the American Neurological Association, the American Neurochemical Society, the Society of Neuroscience, and the American Academy of Neurologists. He received his MD from Yale University.

Myron L. Weisfeldt, MD, Columbia University, is the chairman of the Department of Medicine at Columbia University College of Physicians and Surgeons and the Samuel Bard Professor of Medicine. He is also the director of the Medical Service at the Columbia-Presbyterian Center of the New York Presbyterian Hospital. Before assuming these positions, Dr. Weisfeldt was the Director of the Cardiology Division at Johns Hopkins University School of Medicine. Dr. Weisfeldt received his undergraduate and medical degree from Johns Hopkins University. He received research training at the National Institutes of Health. His clinical training was at Columbia Presbyterian Medical Center, and his cardiology training at the Massachusetts General Hospital. He has served as president of the American Heart Association. Dr. Weisfeldt is a member of the Institute of Medicine, the American Society for Clinical Investigation, the Association of American Physicians, and the Association of Professors of Medicine. He received the Golden Heart Award and the Award of Merit of the American Heart Association.

Kathleen R. Merikangas, PhD, Yale University School of Medicine (liaison to committee from the Board on Neuroscience and Behavioral Health), is professor of epidemiology and psychiatry and director of the Genetic Epidemiology Research Unit at the Yale University School of Medicine. Dr. Merikangas has formal training in clinical psychology, chronic disease epidemiology, and human genetics. She is on the editorial board of several scientific journals and is a member of the Core Scientific Advisory Panel for the MacArthur Foundation Network on Psychopathology and Development and the Psychobiology of Affective Disorders and the Robert Wood Johnson Foundation Research Network on the Etiology of Tobacco Dependence. She has also served on review committees of the National Institute of Mental Health, the National Institute of Drug Abuse, and the National Advisory Mental Health Council Work Group on Mental Disorders Prevention Research and several scientific organizations abroad. Dr. Merikangas has recently joined the National Advisory Council of the National Institute of Drug Abuse. Her major research interests are sources of familial aggregation of psychopathology, comorbidity of mental disorders and substance abuse, vulnerability factors for emotional and behavioral problems in youth, and the public-health impact of prevention.

Workshop Agenda and Speaker Biographies

Stem Cells and the Future of Regenerative Medicine

June 22, 2001
National Academy of Sciences Building
Washington, D.C.

Agenda

OVERVIEW TALKS
8:30-9:30 am

Bert Vogelstein Johns Hopkins University	Opening Remarks
Irving Weissman Stanford University School of Medicine	Overview of Stem Cell Biology
James Thomson University of Wisconsin- Madison	Human Embryonic Stem Cells

STEM CELLS IN DIFFERENT ORGAN SYSTEMS SESSION I
9:30-10:45am

Ernest Beutler The Scripps Research Institute	Bone Marrow Transplantation
Margaret Goodell Texas Medical Center	Stem Cells from Muscle and Bone
Markus Grompe Oregon Health Sciences University	Gene Therapy in the Liver

	Ihor Lemishka Princeton University	Fetal Liver Stem Cells

STEM CELLS IN DIFFERENT ORGAN SYSTEMS SESSION II

11:00am-12:15pm	**Ron McKay** National Institutes of Health	Insulin- Producing Stem Cells
	Iqbal Ahmad University of Nebraska Medical Center	Stem Cells in the Retina
	Fred Gage The Salk Institute for Biological Studies	Repairing the Damaged Brain
	Olle Lindvall Lund University, Sweden	Transplantation of Neural Stem Cells in Humans
12:15-1:15pm	LUNCH	

PUBLIC POLICY PERSPECTIVES SESSION I

1:15-2:30pm	**Thomas Okarma** Geron Group	Biotech Industry and Public Funding
	Arti Rai Washington University Law School	Implications of Restrictions on Stem Cell Research
	Jay Siegel FDA, Office of Therapeutics Research and Review	FDA Perspectives on the Challenges of Stem Cell Therapies

PUBLIC POLICY PERSPECTIVES SESSION II

2:45-4:30pm	**LeRoy Walters** Georgetown University Kennedy Institute of Ethics	Perspectives on Stem Cell Research form Other Countries

	Kevin FitzGerald Georgetown University Medical Center	Arguments Against the Use of Human Embryonic Stem Cells
	David Prentice Indiana State University	Alternatives to Human Embryonic Stem Cells
	George Annas Boston University School of Public Health	Arguments in Favor of the Use of Excess Human Embryos
4:30-5:00pm	**Bert Vogelstein** Summary and Discussion	
5:00pm	ADJOURN	

Audio files from the workshop are available until December 31, 2002, on the Web at http://www.nationalacademies.org/stemcells.

SPEAKER BIOGRAPHIES

Iqbal Ahmad, PhD, is associate professor in the Department of Oph-thalmology at University of Nebraska Medical Center in Omaha. His main research interest is the role of cell-intrinsic and cell-extrinsic factors in maintenance and differentiation of retinal progenitors.

George J. Annas, JD, MPH, is Edward R. Utley Professor and chair, Health Law Department, Boston University Schools of Medicine and Public Health, where he teaches bioethics. He is the author or editor of a dozen books on health law and ethics, including *The Rights of Patients, Judging Medicine, American Health Law, Standard of Care, Some Choice,* and *Health and Human Rights.* He has held a variety of regulatory positions including chair of the Massachusetts Health Facilities Appeals Board, vice-chair of the Massachusetts Board of Registration in Medi-cine, and chair of the Massachusetts Organ Transplant Task Force.

Ernest Beutler, MD, received his degree at the University of Chicago in 1950 and remained at the University of Chicago as house officer and faculty member until 1959 when he became chairman of the Department of Medicine at the City of Hope Medical Center in Duarte, California. In 1979 he assumed the chairmanship of the Department of Molecular and Experimental Medicine at the Scripps Research Institute and position as head of the Division of Hematology/Oncology at the Scripps Clinic. In 1974 while at the City of Hope he initiated one of the early and very successful marrow transplant programs. In 1979 he also orga-nized a marrow transplant program at the Scripps Clinic. Dr. Beutler is editor-in-chief of *Williams Hematology.* He has received the Gairdner Award and has been elected to membership in the American Academy of Arts and Sciences (1975) and the National Academy of Sciences (1976).

Father Kevin T. FitzGerald, SJ, PhD, is the Dr. David Lauler Chair in Catholic Health Care Ethics and associate professor of oncology at Georgetown Medical Center. He received a PhD in molecular genetics and a PhD in bioethics from Georgetown University. His research has focused on the investigation of abnormal gene regulation in cancer and research on ethical issues in human genetics. For the past 10 years he has served as ethics consultant to the National Society of Genetic Counselors. He also serves as a consultant to the Juvenile Diabetes Research Foundation, the United States Catholic Conference, and as a member of the American Association for the Advancement of Science Program of Dialogue on Science, Ethics, and Religion. He is a founding member of Do No Harm: Coalition of Americans for Research Ethics, an organization dedicated to the promotion of scientific research and health care that does no harm to human life.

Fred H. Gage, PhD, is a professor in the Laboratory of Genetics at the Salk Institute and adjunct professor of neurosciences at the University of California at San Diego. Dr. Gage studies regeneration and neurogenesis in the adult brain and spinal cord. He is presently on the National Advisory Council on Aging of the National Institutes of Health and the Advisory Board of the American Society of Gene Therapy. In addition to editorial board duties for a variety of scientific journals, he is chairman of the Scientific Advisory Board of the Christopher Reeve Paralysis Foundation. He is the recipient of several research awards including the Christopher Reeve Second Annual Medal Award, the Mathilde Solowey Lecture Award in Neuroscience, the Robert J. and Claire Pasarow Foundation Award, the Max Planck Research Award, the Theobald-Smith Award, and the Bass Foundation Lecture Award.

Margaret Goodell, PhD, is an assistant professor in the Center for Cell and Gene Therapy Departments of Pediatrics, Molecular and Human Genetics, and Microbiology and Immunology at Baylor College of Medicine. Dr. Goodell has worked on stem cells derived from adult tissues for over 10 years, first focusing on those in the hematopoietic system and more recently from a number of other tissues. Her work has indicated that adult stem cells (mesenchymal stem cells) can differentiate in bone, cartilage, and brain cells (astrocytes) in culture.

Marcus Grompe, MD, PhD, is a pediatrician in the Department of Molecular and Medical Genetics, Oregon Health Sciences University. Using a mouse model of hereditary tyrosinemia (a genetic disease that is associated with severe liver deficiency in infants), his laboratory has found that more than 90% of host hepatocytes can be replaced by a small number of transplanted donor cells in a process called "therapeutic liver repopulation," which is analogous to repopulation of the hematopoietic system after bone marrow transplantation.

Ihor Lemischka, PhD, is professor in the Department of Genetics, Genomics, and Bioinformatics at Princeton University. His research analyzes hemopoietic differentiation using retroviruses as markers and has focused on gaining insight into the in vivo clonal behavior of the most primitive fetal liver or adult bone marrow hematopoietic stem cells. In particular, his laboratory is interested in understanding the mechanistic aspects of: (1) self-renewal vs. commitment decisions during stem cell proliferation and (2) the nature of commitment decisions as they partition the complete set of developmental potential into subsets or, in other words, the establishment of the primitive portion of the hematopoietic hierarchy.

Olle Lindvall, MD, PhD, is professor of neurology and chairman of the Department of Clinical Neuroscience, Lund University, Lund, Sweden. Dr. Lindvall's current main research interests are the use of cell and gene therapy for preservation and restoration of function in acute and chronic brain diseases. Since 1983 he has been in charge of the clinical cell transplantation program for patients with Parkinson's disease at Lund University.

Ron McKay, MD, is the chief of the Laboratory of Molecular Biology at the National Institute of Neurological Disorders and Stroke. Dr. McKay has made major contributions to the identification of stem cells in the nervous system. His group is developing cell therapies for diabetes, neurological, and cardiac disease.

Thomas Okarma, MD, PhD, is president and chief executive officer of Geron Corporation. After receiving his PhD and MD degrees at Stanford University, Dr. Okarma became a member of the faculty in the Department of Medicine at Stanford University School of Medicine in 1980. Dr. Okarma left Stanford in 1985 and founded Applied Immune Sciences, Inc. (AIS) where he was chief executive officer. By the time of its acquisition by Rhone-Poulenc Rorer in 1995, AIS was in advanced cell therapy human clinical trials in cancer and bone marrow transplantation and in early gene therapy human trials in breast cancer.

David A. Prentice, PhD, is professor of life sciences at Indiana State University, adjunct professor of Medical and Molecular Genetics for Indiana University School of Medicine, and a founding member of Do No Harm: The Coalition of Americans for Research Ethics, an organization dedicated to the promotion of scientific research and health care that does no harm to human life. One current focus of his research is on adult stem cells and their differentiation signals.

Arti Rai, JD, is assistant professor of law at the University of Pennsylvania School of Law. She attended Harvard Medical School prior to receiving her law degree from Harvard Law School. Professor Rai teaches and writes in the areas of biotechnology and the law, patent law, and health care regulation. Her recent work addresses the interaction between the public and private sectors in biomedical research. She is a co-author of *Law and the Mental Health System* (West Publishing) and serves on the Board of Editors of the *American Journal of Law and Medicine*.

Jay Siegel, MD, is director of the Office of Therapeutics Research and Review (OTRR) at the Center for Biologics Evaluation and Research (CBER), Food and Drug Administration (FDA). His office has responsibility for regulation of biological therapeutics, including cell therapies, gene therapies, monoclonal antibodies, cytokines, and other proteins. This office has over a decade of experience in applications review, research, and development of scientific standards and policy with regard to hematopoietic stem cell related products and, in 2000, convened the first FDA advisory committee conference on neurologic stem cell products. Since joining CBER in 1982, Dr. Siegel has served as founding director of the Division of Clinical Trial Design and Analysis, deputy director of the Division of Cytokine Biology, chief of the Laboratory of Cellular Immunology, and senior investigator in the Division of Virology. He trained in medicine and infectious diseases at Stanford University School of Medicine, in internal medicine at the University of California, San Francisco, and in biological sciences at the California Institute of Technology.

James A. Thomson, DVM, PhD, is a University of Wisconsin-Madison developmental biologist in the Department of Anatomy in the School of Medicine and also serves as the chief pathologist at the Wisconsin Regional Primate Research Center on the UW-Madison campus. Dr. Thomson received his doctorate in veterinary medicine in 1985 and his doctorate in molecular biology in 1988, both at the University of Pennsylvania. Since joining the Wisconsin Regional Primate Research Center, he has conducted work on the isolation and culture of non-human primate and human embryonic stem cells, undifferentiated cells that have the ability to become any of the cells that make up the tissues of the body. Dr. Thomson directed the group that reported the first isolation of embryonic stem cell lines from a non-human primate in 1995, work that led his group to the first successful isolation of human embryonic stem cell lines in 1998. Dr. Thomson is the scientific director of the WiCell Research Institute, a private subsidiary established by the Wisconsin Alumni Research Foundation to supply cells to support research for both academic and non-academic researchers.

LeRoy Walters, PhD, is the Joseph P. Kennedy Professor of Christian Ethics at the Kennedy Institute of Ethics and professor of philosophy at Georgetown University. He chaired the Recombinant DNA Advisory Committee from 1993 to 1996. He is the author of *The Ethics of Human*

Gene Therapy (1996) and co-editor of *Source Book in Bioethics: A Documentary History* (2000).

Irving L. Weissman, MD, is Karel and Avice Beekhuis Professor of Cancer Biology, Cell and Developmental Biology at Stanford University. His research encompasses the phylogeny and developmental biology of the cells that make up the blood forming and immune systems. His laboratory has identified and isolated the blood-forming stem cell from mice and has defined, by lineage analysis, the stages of development between the stem cells and mature progeny. In addition, the Weissman laboratory has pioneered the study of the genes and proteins involved in cell adhesion events required for lymphocyte homing to lymphoid organs in vivo, either as a normal function or as events involved in malignant leukemic metastases. Dr. Weissman has been elected to the National Academy of Sciences and to the American Association for the Advancement of Science. He has received the Kaiser Award for Excellence in Preclinical Teaching, the Pasarow Award, and the Outstanding Investigator Award from the National Institutes of Health.

Index

A

Adipocytes, 15, 28, 32
Adult stem cells, 19-29, 41, 55
    animal models, 21, 23, 25-26, 27, 28-
       29, 47
    bone marrow, 2, 7, 16, 19, 21, 41, 58
    brain, 2, 12, 16, 23, 24, 25, 27-28
    cancer, 22, 24
    cardiovascular system, 21, 28
    cloning and, 11
    committee recommendations, 5, 55, 56
    defined, 16-17, 67
    differentiation, 2, 7-8, 16-17, 20, 21,
       27-28, 29
    embryonic stem cells *vs*, 2-3, 17, 56
    eyes, 23, 25, 25-26, 28
    hematopoietic stem cells (HSCs), 16,
       19-23, 25, 26, 29, 47
    hepatic system, 23, 26
    histocompatibility, 21-22, 24, 26
    multipotent, 28
    muscle, 16, 21, 23, 25, 28
    neural, 16, 21, 23, 24, 26, 27, 28
    pancreatic, 23, 26
    philosophical, ethical, and legal issues,
       47
    plasticity, 16, 24, 25, 26, 28, 41, 47
    proteins, 12, 22, 26-27
    skin, 21, 23, 27, 33, 41
Alzheimer's disease, 8
Animal models, 2
    adult stem cells, 21, 23, 25-26, 27, 28-
       29, 47
    bone marrow transplants, 21

    committee findings and
       recommendations, 55-56
    diabetes, fetal stem cells, 16
    embryonic stem cells, 32-34, 35, 36
    hematopoietic stem cells, 21, 23, 25, 29
    neural stem cells, 13, 27
Antigens
    see also Histocompatibility
    defined, 67, 71
    surface antigen, 22-23, 38, 71, 72
Attitudes and beliefs, *see* Philosophical,
    ethical and legal issues; Public
    opinion
Autoimmune diseases, 8
    diabetes, 8, 28, 35, 84
Autologous transplants, 24, 27
    defined, 67

B

B cells, 20
Birth defects, 8
Blastocysts, 13, 14, 31, 32, 37
    cloning, 11
    defined, 67
    inner cell mass, 15, 31, 37, 70; *see also*
      Embryonic stem cells
Blood, *see* Hematopoietic stem cells
Bone cells, 16, 28, 79
Bone marrow stem cells
    see also Hematopoietic stem cells
    adult, 2, 7, 16, 19, 21, 41, 58
    committee workshop agenda, 79
    defined, 67-68

O

Osteoporosis, 8
Ovum, 13, 14, 15, 37
    cloning, 11
    defined, 71

P

Pancreas
    adult stem cells, 23, 26
    committee workshop agenda, 80
    diabetes, 8, 16, 28, 35, 84
    embryonic stem cells, 32, 34-35, 36
    fetal stem cells, 13, 16
Parkinson's disease, 8, 13, 27, 35, 84
Pharmaceuticals, *see* Drugs,
    immunosuppressive
Philosophical, ethical and legal issues, *xi-*
    xii, 1, 3-5, 10, 12, 43-48, 50, 51-53,
    55, 57-58, 85, 86-87
    adult stem cell research, 47
    committee workshop agenda, 80-81
    DNA research, 3-4, 51-53, 58
    international perspectives, 46, 80
    public opinion, *xi*, 1, 52-53, 55, 57
    religious, 44-45
Placenta, 16
    defined, 71
Plasticity
    see also Multipotent stem cells
    adult stem cells, 16, 24, 25, 26, 28, 41,
    47
    defined, 12, 71
    embryonic stem cells, 41
Pluripotent stem cells, *see* Multipotent
    stem cells
Private sector, 3, 49-52, 57, 85
    committee workshop agenda, 80
Progenitor cells
    hematopoietic, 20
    pancreatic, 13
Proteins, adult stem cells, 12, 22, 26-27
Public opinion, *xi*, 1, 52-53, 55, 57
    religious issues, 44-45

R

Religious, 44-45
Renal system
    embryonic stem cells, 34-35
Reproductive system
    see also terms beginning "Embryo," and
    "Fetal"
    cloning, 10, 11, 48
    in vitro fertilization, 43, 48, 70
    ovum, 11, 13, 14, 15, 37, 71
    placenta, 16, 71
    umbilical cord, 16, 22
    zygotes, 13, 14, 15, 37, 72

S

SCNT, *see* Somatic cell nuclear transfer
    technique
Skin cells, 7, 13, 16
    adult stem cells, 21, 23, 27, 33, 41
    burns, 8
    embryonic stem cells, 16, 33
Smooth muscle cells, 15
Somatic cell nuclear transfer (SCNT), 5,
    10, 11, 12, 37, 39, 45, 46, 47, 59
Somatic cells, general, 29
    committee recommendations, 5
    defined, 71
Spinal-cord injuries, 8, 83
Stromal cells, 33
    defined, 71
Surface antigens, 22-23, 38, 71
    defined, 72

T

T cells
    defined, 72
    hematopoietic stem cells, 20, 21
Trophoblasts, 13

U

Umbilical cord, 16, 22
Undifferentiated cells, *see* Differentiation
United Kingdom, 46

W

White blood cells
  see also Hematopoietic stem cells;
    Lymphocytes; T cells

defined, 72
World Wide Web, *see* Internet

Z

Zygotes, 13, 14, 15, 37
  defined, 72